Scanning:
Your Personal
Consultant

Scanning: Your Personal Consultant

Jonathan Hornstein

Ziff-Davis Press
Emeryville, California

Development Editor	Valerie Haynes Perry
Technical Reviewer	Bruce Fraser
Project Coordinator	Cort Day
Proofreader	Carol Burbo
Cover Illustration and Design	Regan Honda
Book Design	Regan Honda
Technical Illustration	Sarah Ishida and Dave Feasey
Word Processing	Howard Blechman
Page Layout	M.D. Barrera
Indexer	Mark Kmetzko
Cover copywriter	Sean Kelly

Ziff-Davis Press books are produced on a Macintosh computer system with the following applications: FrameMaker®, Microsoft® Word, QuarkXPress®, Adobe Illustrator®, Adobe Photoshop®, Adobe Streamline™, MacLink®Plus, Aldus® FreeHand™, Collage Plus™.

If you have comments or questions or would like to receive a free catalog, call or write:
Ziff-Davis Press
5903 Christie Avenue
Emeryville, CA 94608
1-800-688-0448

ISBN 1-56276-297-4

Manufactured in the United States of America
10 9 8 7 6 5 4 3 2 1

To Patrice…for being there.

TABLE OF CONTENTS

Introduction
x

CHAPTER 1
The Benefits of Doing It Yourself
1

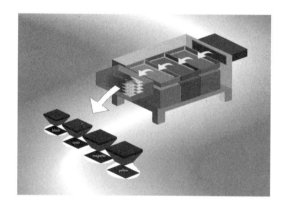

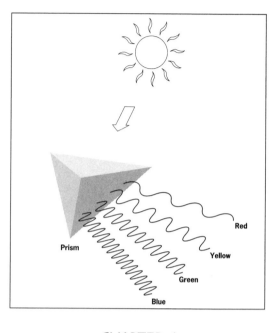

CHAPTER 2
The Production Process
8

CHAPTER 3
The Types of Scanners
22

CHAPTER 4
Resolution
40

CHAPTER 5
Step-by-Step Scanning
54

CHAPTER 6
Understanding and Using Color
74

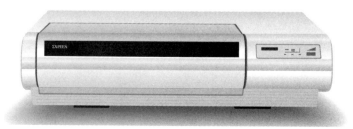

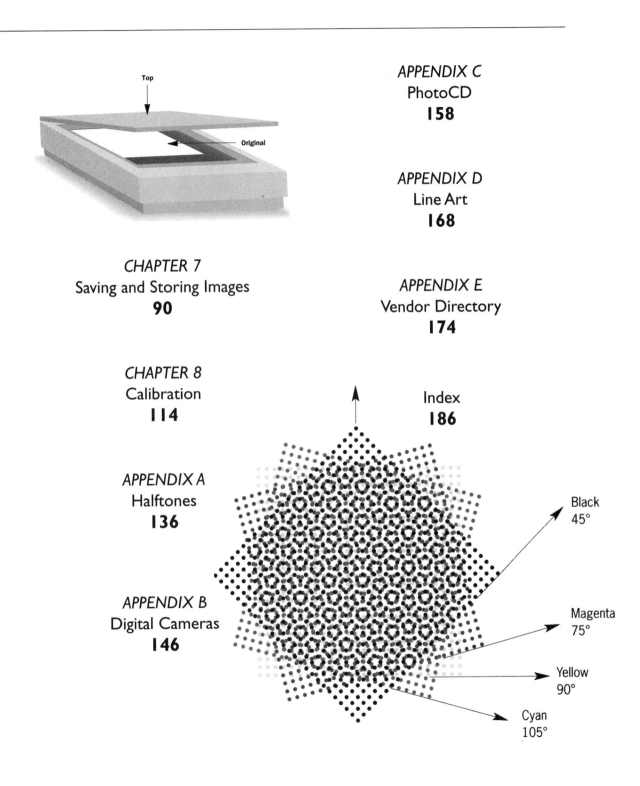

APPENDIX C
PhotoCD
158

APPENDIX D
Line Art
168

CHAPTER 7
Saving and Storing Images
90

APPENDIX E
Vendor Directory
174

CHAPTER 8
Calibration
114

Index
186

APPENDIX A
Halftones
136

APPENDIX B
Digital Cameras
146

Top

Original

Black
45°

Magenta
75°

Yellow
90°

Cyan
105°

ACKNOWLEDGMENTS

I'm deeply indebted to Bruce Fraser, the patient and diligent technical reviewer for this book, who provided invaluable input. Thanks also to Erik Holsinger for letting me share the benefit of his experience as an author.

I would also like to express my appreciation to the following people for the opportunities they've given me to work, teach, and learn: Paul Raybin of Creative Communications; Steve Schaffran of Techmark; Henry Brimmer of Photo Metro; Steve Whaley of the Graphic Arts Institute; Dan Gonzales; Dennis DeSantis of the Academy of Art College; and Dean Fernandes of Landor Associates.

Any acknowledgment would be incomplete without thanking the all folks in the photo department at *U.S. News & World Report* who helped to make my first experience in publishing such a positive one. There were many of you, but I'd especially like to thank Richard Folkers of *U.S. News & World Report*, John Echave of *National Geographic*, and Steve Larson of the *Denver Post*. Thanks also to Ellie Leishman of *MacWEEK* for taking a chance and hiring a photo editor who didn't know RAM from a RIP.

Finally, I would like to thank my mom and the rest of my family for supporting me in all of my various endeavors over the years. And special thanks to Maggie for raising me right.

INTRODUCTION

As recently as six years ago, scanning color photographs and creating separations were the exclusive domain of the "color priesthood" at prepress and printing trade shops. These folks had years of color training and access to high-end, and high priced equipment. If you wanted to reproduce color images in print, you had to pay their hefty prices and do it on their schedule.

Today, you can buy your own scanner for less than $1,000. Add the cost of a computer system and image editing software, and you have all the tools you need for doing your own scanning and separations. Why, then, are color trade shops still in business? Because there is a difference between simply having tools, and having the skill, knowledge, and experience to use them properly.

While having access to tools does not automatically bestow the skill required to use them, it does affix responsibility. One of the biggest challenges facing designers, photographers, and publishers today is the fact that the tools used in desktop publishing and imaging dictate that those professionals take responsibility for parts of the production process of which they have little knowledge and even less experience. When print production was practiced only by skilled tradesmen, the lines of responsibility were clearly defined by the tools, training, and almost a century of accepted practice. Since the introduction of desktop publishing, roles and responsibilities have been in flux. No matter how the situation evolves, those involved in the creative side of publishing will become increasingly responsible for the production side as well.

If you've been disappointed with your attempts at scanning, don't be discouraged. While it's easy enough to physically scan a photograph, obtaining satisfactory, predictable results requires knowledge and experience. But most users of desktop scanners don't aspire to be full-time scanner operators, they simply want scanning to help them do their primary job better, faster and cheaper.

The good news is that in most cases, you don't need to be a professional scanner operator or use high-end equipment to get satisfying results. Armed with some basic knowledge about color, scanning, proofing, and printing even a relative newcomer to desktop scanning using inexpensive equipment can produce acceptable, professional-looking images. As with any endeavor, patience and a willingness to learn are the keys to success. The time and effort you invest in learning how to do your own scanning will be paid back many times over by savings in time and money, while giving you more control over your work.

CHAPTER

The Benefits of Doing It Yourself

The Challenges
The Rewards

The attraction of desktop publishing is undeniable. As professional designers and publishers, desktop publishing offers us the promise of saving time and money while allowing greater control over our work. Turnaround times are faster because we don't have to send work to outside vendors. Costs are lower because we can do the work ourselves on inexpensive desktop equipment, rather than paying others to do it for us on expensive, proprietary systems. And doing the production ourselves gives us direct control over how our work will look in print, instead of having our vision interpreted by others.

Unfortunately, reality rarely lives up to this ideal. Too often both quality and productivity fall off dramatically when we take on production work ourselves. Unfathomable technical problems slow our work down to a crawl, or stop it altogether. We end up spending more money because problems with quality and productivity cause us to come up against deadlines, and we're forced to pay rush charges to outside vendors to get the work done on time. And the control we desperately want turns into a curse when after repeated attempts the image or page still doesn't look good or won't print at all.

In no other area of desktop publishing are the promises more alluring or the challenges more daunting than with scanning. Aside from the actual printing bill, sending art and photographs out to be scanned at a color trade shop can be one of the most expensive items in a publication's production budget. Sending the work out also means tighter production deadlines in order to avoid the rush charges required for fast turnaround. Both artistic and editorial flexibility are limited because art and photographs must be chosen earlier in the production process. And the trade shop's scanner operator decides how the photo looks on the printed page; you don't get to make the final decision. All of these factors conspire to make doing your own scanning an attractive option.

The Challenges

Creating your own professional quality scans is one of the great challenges of desktop publishing, and the potential pitfalls are many. These include (but are by no means limited to) scanning photos at the wrong resolution; using incorrect color settings; under- or over-sharpening; basing color judgments on an uncalibrated monitor or proof; or choosing a poor original to begin with. These dangers exist in addition to the technical problems we may face when setting up and using a personal computer with the necessary software and peripherals.

On top of all this, handling the large, complex files which result from scanning can cause performance and storage problems. And if the quality is not acceptable when the images appear in print, the attempt to save time and money will have ended up wasting both.

Until recently, scanning color photographs required a professional scanner operator working on a system costing over $100,000. Today anyone who is at least somewhat familiar with personal computers and can plunk down $1,000 to buy an inexpensive desktop scanner or $35/hr. to rent one can do a scan. While in both cases a photograph is scanned, the results can be—and often are—very different. Anyone with access to a desktop scanner can physically scan a photograph, but that does not guarantee that the printed image will be either accurate or pleasing. Unless you are armed with knowledge of basic issues such as resolution, calibration, color gamuts and file formats, disappointing results are almost inevitable.

Color trade shops use expensive, sophisticated scanners designed for high-quality, high-volume production. But what largely accounts for the high-quality of trade shop work is not the scanner but the scanner operator. Professional scanner operators have years of training and experience doing just one thing—scanning images. Skilled in evaluating originals, finely tuning complex scanners, and managing production workflow, they are also knowledgeable about photography, proofing and offset printing. They are highly respected and well paid professionals. Even the most sophisticated software cannot totally replace these skilled and experienced workers.

You can learn a lot about scanning and separation when you work with a color trade shop, but there's a big difference between learning *about* scanning and learning *how* to do it yourself.

You don't need to develop the level of skill and knowldege of professional scanning operators in order to get satisfactory results. But as with any valuable skill, there is a learning curve. While you can learn both the basics and important details about scanning from books and software manuals, there's still no substitute for experience. Trial and error are invaluable learning tools. The key is to be methodical and willing to refer to books and manuals as references.

The Rewards

Despite the challenge of doing your own scanning, the benefits to be reaped are significant. By investing in a scanner and spending the time to learn how to use it properly, you'll be able to

- Save time

- Save money

- Gain control over your work

The following section covers these benefits in more detail.

Saving Money

It can cost $75 or more to send a photograph to a trade shop for separation. There are other costs involved in using a trade shop, such as messengers and sales tax (in some states), and in doing your own scanning, such as time on a host computer while scanning and non-allocable expenses such as office space and electricity. The cost of film and proofs also needs to be added to calculate the full price of producing scans and separations done either in-house or at a trade shop.

Doing the scan yourself may take as little as 10 minutes of your time, plus of course the initial cost of the equipment. This lowers your overall production costs, leaving more money for other items in your budget. And by doing your own scanning, you can use many more photos and still save money. Because you don't need to pay a color trade shop for each image you use, the production cost of using additional photos is marginal.

See the sidebar, "In-House Scanning: A Case Study," at the end of this chapter for an example of how you can realize savings.

Saving Time

Keeping scanning in-house makes your scans available much more quickly, while avoiding the surcharges required by most trade shops for a turnaround that is under two days. Even if you are willing to pay the typical 200%–300% surcharge required by most color shops for ASAP turnaround, you still need to take into account the additional time that is involved. For example, a messenger needs time to deliver your original to the trade shop and then return the scan to you, and it also takes time to log the job in and out on both ends. The whole process can easily take over an hour, and that doesn't include the time it takes to actually do the scan.

Doing your own scanning can push back production deadlines, allowing more time for the creative work. Doing it yourself can also extend the deadlines for printing time-sensitive information, such as late-breaking news. No matter how good a photo or piece of art is, it's worthless to you unless it's ready in time to meet your deadline.

Gaining Control

Finally, doing your own scanning gives you more control over how the picture will look in print. With good software and a bit of experience, you can learn color and tone correction, and even color enhancement. No scanner, not even expensive drum scanners, can capture all of the detail and color information found in a transparency. The scanner operator has to make decisions about what is important in the image, such as the color of a particular object or detail in a specific area, and try to reproduce that information as faithfully as possible. That part of the image which is deemed to be less important may not match the original at all. While you can try and communicate to the scanner operator what you think is most important and should be preserved, the surest way to get what you want is to do it yourself.

Conclusion

Desktop scanning can save you time and money, while giving you more control over your work. Unfortunately it also has the potential to waste both time and money, while giving you unsatisfactory results. The first step in reaping the significant rewards of doing your own scanning is to understand the challenges it presents and invest some time in learning how to overcome them.

In-House Scanning: A Case Study

In order to see the potential benefits of desktop scanning, let's take the example of a fictitious publication, Gardening Today. This weekly trade publication uses 20 color photographs per issue. Sending them out to a color trade shop costs $75 per image, or $1,500 per week for all 20 photographs. Deciding to bring their scanning in-house, the art department purchases a desktop scanner for $2,500. With approximately one-half hour needed to scan, separate and proof each image, it would take about 10 hours of a worker's time to process the photos for each issue. If that worker made $25 per hour, the labor cost would be $250 per week. If the scanner is conservatively estimated to last two years (or 100 issues), its cost would come to $25 per week. By switching to desktop scanning, Gardening Today would realize over $1,200 in savings per week, or over $60,000 per year.

Before they brought scanning in-house, the art staff needed to choose the photos to be used almost three days before press time. With the lead time needed to get the photographs, the editors were forced to make story decisions over a week before the readers received the magazine. With desktop scanning, Gardening Today can push their production deadline back two days. This allows them to include all the late-breaking gardening news, making their publication more timely and vital.

Gardening Today can also be a better-looking magazine. The art staff has more time to design each issue because the lead time for production is shorter. And the marginal production cost of using photographs is small, so they can use more photographs without being concerned about overspending their art budget.

Table 1.1   **Cost of Color Trade Shop vs. In-House Scanning**

	Trade Shop	In-House
Number of images	20	20
Cost per scan	$75	$13.75
Cost per issue	$1,500	$275
Cost per year	$75,000	$13,750

Prices for both a trade shop and in-house scanning do not include the costs of film and proofs.

C H A P T E R

The Production Process

Scanning in Context

Types of Output

The Production Process from A to Z

Correcting Mistakes

Print production is an integrated process of which scanning is just one step. Knowing what happens before and after we scan tells us what we need to know to get satisfying results. In this chapter, we look at scanning in the context of the entire production process.

Scanning in Context

Unless we are simply experimenting, every photograph we scan has a purpose. I'm referring to the physical form the image will take once it's made into a final product, such as a magazine, brochure, package or even a video presentation. Keeping this in mind, scans need to be made specifically for the reproduction method used to make the final product. Because there is no such thing as a universally ideal scan, a scan can only be judged by how it looks after producing the magazine, brochure, and so on.

With scanning, context is everything. We can't really understand how to make good scans without knowing something about the entire production process. Following the steps in the production process and understanding how each one will be carried out determines how the final product will be manufactured.

Starting at the End

The principle of "Starting at the End" is the foundation of any production process, and publishing and imaging are no exceptions. "Starting at the End" is a plan that tells you where you are going, which in turn helps to guide you every step of the way. When scanning and adjusting images, you're making important decisions about how the image will look when reproduced. If you don't know the image's destination, or final output, you won't have enough information to make the right choices during the production process.

Types of Output

Every type of output device has its own unique characteristics. Each differs in the range of colors it can reproduce, the degree of contrast it allows, and how precisely it registers colors, just to name a few distinctions. Without taking these factors into account, your scans won't look as good as they could, and may not look good at all.

The following sections discuss some of the most common destinations for scans.

Offset Printing

Most of the magazines, brochures, and packages we see are printed on an offset press. In offset printing, individual dots of cyan, magenta, yellow and black (CMYK)—the primary printing colors—are placed onto the paper in such a way that we perceive the whole visual range of colors.

These presses come in two distinct types: web presses and sheetfed presses. *Web presses* print on large rolls of paper at very high speeds. Most magazine and large run print jobs are printed on a web press. The other type is the *sheetfed press*. As the name implies, these presses print on individual sheets of paper. Sheetfed presses are slower than webs, but they are generally capable of producing higher quality work.

Color Printers

Color printers make a color print directly from your digital file. There are many different types of color digital printers available, and they vary greatly in price, quality, and the size of their output. Some, like dye-diffusion printers (also called dye-sublimation printers) and inkjets, are capable of near photographic quality reproduction. Others, like thermal wax and solid wax printers, reproduce type and solid colors better than photographs. Color laser printers and their cousins, color copiers with computer interfaces, produce both images and type equally well. A print from one of these devices may be used as the final product, or it may be used as an intermediary proof for an image which will ultimately be printed on an offset press.

Video

A scanned image may not be printed at all, but instead used on screen in a digital slide presentation, game, multimedia project, or it may be transferred to videotape. Scanning for video output is generally easier than scanning for print reproduction because the image quality of video is more limited. Video is also much less sensitive to changes in color, contrast, and resolution than are most printing methods. Precise color matching is impossible with most types of video output because the characteristics of video monitors vary widely.

The Production Process from A to Z

Producing an ad, publication or presentation is a complex process that consists of many steps. Each step affects all the other steps as well as the final output, so everything must be carefully coordinated. Scanning is no exception. In order to create quality scans, you need to understand how each step of the process is integrated with the others. Without this knowledge, it is nearly impossible to make quality scans.

Following is an overview of the entire print production process from beginning to end. The workflow and the tools used will vary depending upon the type of project, the final output, and the budget. Figure 2.1 illustrates what is involved.

Figure 2.1
The
production
process

Slide scanner

Color printer

Figure 2.1
The
production
process
(Continued)

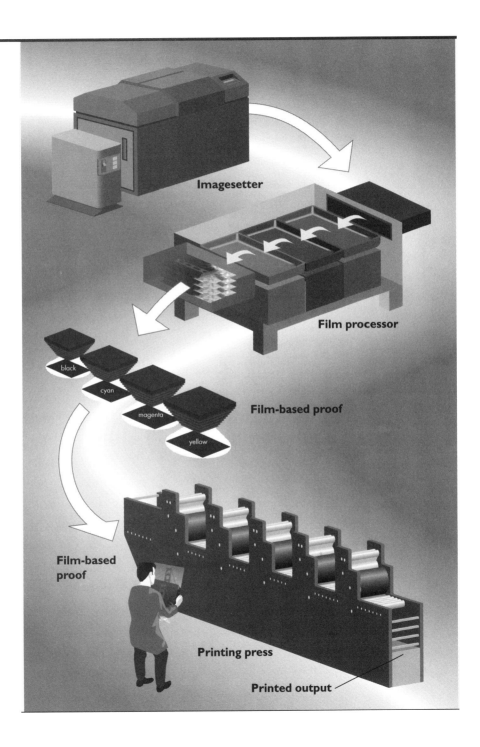

Imagesetter

Film processor

black
cyan
magenta
yellow

Film-based proof

Film-based
proof

Printing press

Printed output

Before You Start

Before you can start scanning you must ask and answer the following questions:

What type of device will be used for final output? Knowing what the final output device will be tells you several important things. The first is the resolution at which the image must be scanned. Photos must be scanned in at sufficient resolution in order not to appear coarse or blurry. (Resolution is the topic of Chapter 4.) The second is the range of colors and tones the output device can reproduce. This information will guide you in setting the tone and color correction controls when you are scanning.

When scanning for output to an offset press, you'll need to know what type of press will be used, the line screen (the frequency of halftone dots) at which the image will be printed, and the amount of dot gain (the amount the ink will spread when it hits the paper) that is expected. If you are printing to a digital color printer, you'll need to know what resolution the device requires.

At what size will the picture be used? It's important to determine the picture's final size before it's scanned. Without this information, the scan will either be made at too low a resolution, resulting in poor image quality, or too high a resolution, producing an unnecessarily large file which is cumbersome to work with, difficult to store. and slow to print.

How will the image be integrated into the next step in the process? This factor will determine the format in which your scan should be saved, and whether compression should be used. For example, a page layout program may require imported images to be in a specific format in order for them to separate and print properly. A presentation program may have different requirements. Various output devices may print some file formats faster than others. If the scanned image is to be sent via modem, a compressed format should be chosen. (File formats and compression are discussed in greater detail in Chapter 7, "Saving and Storing Images.")

The Steps of Scanning

Chapter 5 will take you through the scanning process in greater detail. In the meantime, here is a brief overview, putting it in the context of the entire production process.

The first step in scanning (after ascertaining the important information needed about the final output) is selecting the photograph to be scanned. After choosing an image based on its design or editorial content, it needs to be inspected from a production point of view. The original is first checked for dust, scratches, and fingerprints. Assuming there are no serious problems, any loose dirt is cleaned off.

After cleaning, the original is mounted in the scanner and the focus is set. How this is done depends on the type of scanner. For example, flatbed scanners don't require focusing because the original and the part of the scanner which reads the image is always at a fixed distance.

Prescanning is the next step. A *prescan* is a quick, low-resolution scan which allows you to see a preview of your image for cropping purposes. If the scanner and software allow, you can also make contrast and color adjustment based on how the prescan looks on the screen. Then the actual scanning is done. Once the image is scanned, adjustments are made to the tone, color balance, and contrast.

Next the image is sharpened. It may sound as if sharpening is needed to compensate for a blurry original or out-of-focus scan, but it is in fact a crucial step in reproducing any image. Sharpening selectively adds contrast to an image, making it appear much more vibrant on the printed page. Sharpening can only restore the sharpness lost during the scanning process. It cannot bring an out-of-focus original into focus.

Finally, the image is saved in the correct file format. The format chosen should be based on the requirements (if any) of the application into which the image will be imported and of the final output device. The file may also need to be compressed if it will be transmitted by modem or needs to fit on a floppy disk or removable medium.

Separations

The picture will need to be separated if it will be printed on an offset press, or on most color printers. *Separating* an image means changing its component colors from the additive colors, red, green, and blue (RGB) used by scanners and monitors, to the subtractive colors cyan, magenta, yellow, and black (CMYK), which are used for printing. As in all other areas of reproducing an image, the final output device will determine how the image will be separated. The type of image, its contrast range, and the amount of neutral tones present influence the conversion from additive to subtractive color.

No matter what printer or press is used for the final output, there will be some loss of color fidelity from the original photograph. Photographic film and scanners can capture a broader range of colors than can be reproduced with ink, toner, or dye on paper. A good separation will preserve the color fidelity of the most important parts of the image. The rest of the image will often need to be compromised.

Integrating Images

If the scan has been prepared correctly (at the correct physical size, resolution, and file format) then bringing the image into a desktop publishing or multimedia application is a straightforward process. Simply using the application's Get Picture, Import, or Place command will integrate the scan into the page or presentation.

Once you import a scanned image into another application, the edges should be examined closely to make sure there are no gaps between the edge of the image and any border or background color. What may be hard to see on your monitor will be readily apparent on printed output.

Beware of changing the size of a photo once it's been brought into another application. Making an image larger, or even smaller after it's been imported can severely degrade the quality of the image. (You will find more on resizing images in Chapter 4, "Resolution.")

Using the Same Scan in Several Projects

Q: I want to use the same photograph in an offset printed brochure, a poster output on a digital printer, and in a multimedia presentation. Can I use the same scan for each of these projects?

A: To get the best possible reproduction in all three of the products, the image should be scanned three separate times, each time taking into account the special characteristics of each type of final output. In a pinch, you can use one of the scans for your printed output in your multimedia presentation because video monitors can reproduce a wider range of colors, and therefore will be more forgiving when displaying an image. But the images won't look as good as they would if you had scanned them specifically for display on a monitor. As the use of color-management systems becomes more prevalent, it will become possible to use one scan for many different types of output. (For more on color-management systems, see Chapter 8, "Calibration.")

Digital Color Proofing

A digital print from a color printer may be your final output. For example, if you want to scan several images, make a composite of them, and make a print to hang on your wall, then a digital print will be the final product. However, if an offset press will be used for the final output, you may use the digital print as a way of seeing what the image will look like after it's printed on a press.

When using a printer that can simulate the output of an offset press, it is still recommended that a film-based proof be made. (Film-based proofs are covered later in this chapter.) Sometimes, however, the budget dictates that the digital color print be the final proof before the image goes to press.

Even if the color printer you use is not capable of letting you preview how your image will look once it's printed on an offset press, it is still wise to proof the image on a color printer. Producing an un-calibrated proof will still show some things that cannot be seen on a monitor, such as problems with resolution, file format, and image placement.

Relying on Digital Proofs

Whether or not you can rely on a digital proof as a preview of what the image will look like when printed on an offset press is subject to much debate. Digital color proofs are generally less expensive than film-based proofs and therefore an attractive alternative. Some devices, such as the Rainbow dye-diffusion printer from 3M and the Iris inkjet printer are specifically designed to be used as proofing devices. There are also several high-end digital proofing methods, such as the Kodak Approval System. They can closely match printed output but are quite expensive. Any color proof, either digital or film-based, can only simulate (but not exactly match) what printed output will look like because each device uses a different method of creating colors.

In order to be able to use a digital color proof as a press proof, you not only need to use a color printer capable of simulating output from an offset press, but the printer also needs to have a calibration setting for the particular type of press your job will be printed on. (See Chapter 8, "Calibration.") Don't ever assume that using a digital proof is okay without consulting your print shop.

Printing to Film

If the final output will be printed on an offset press, the next step is to print the film separations to an imagesetter. An *imagesetter* works much like a black-and-white laser printer, except it can "print" on film and at very high resolution. It does not in fact print on film, but rather uses laser light to expose the film. Standard photographic chemistry is then used to process the film.

A color photo is printed to the imagesetter as separate pieces of film for each of the four process colors cyan, magenta, yellow, and black (CMYK). Black-and-white film is used for each of the four plates. Here's what happens if the film is printed as a negative when the ink is applied to the paper on the press: The clear areas on the film will be covered with the full amount of the appropriate color; the black areas will be paper white; and the gray areas will have varying amounts of color, depending on their shade.

Relatively new technologies are evolving called Direct-to-Plate and Direct-to-Press printing. These processes do away with the intermediary step of producing film separations and instead send the digital files either directly to a plate making device, or straight to the press. One of the great challenges in Direct-to-Plate and Direct-to-Press printing is getting an accurate proof without the use of film.

Film-Based Proofs

Once the film separations have been printed, a color proof can be made directly from the film. Even if you've used a digital proofing method designed to simulate printed output, it is still recommended that a film-based proof be made. These proofs give a better representation of how the printed image will look than a digital print can. And because these proofs are made directly from the film that will be used to make the printing plate, they can help you detect flaws on the film such as scratches, poor exposure, and improper development.

Film-based proofs fall into two general categories: laminated and non-laminated (also known as "overlay"). *Laminated proofs* usually offer better color fidelity and are better at showing dot patterns than non-laminated proofs. They also tend to cost substantially more. Some brand names of laminated proofs include Chromalin, PressMatch, and Matchprint. Non-laminated brands include Color Key and Chromacheck.

On the Press

Once the film-based proof is approved, the plates for the printing press can be made. The plates themselves are flexible metal sheets covered with a light sensitive emulsion much like photographic film. The plates are made by placing the film on top of the plate material before exposing it to a special light source. This leaves areas on the plate, corresponding to the image, to which ink adheres, while water is used to repel ink on the rest of the plate. For color images, four plates are made, one for each of the process printing colors, cyan, magenta, yellow, and black (CMYK).

The plates are then mounted on the press by wrapping them around a drum. When the press starts, each plate rolls through a trough filled with the color of ink that plate will reproduce. The ink-laden plate then rolls in contact with an ink transfer drum known as a *blanket*. Finally, the blanket rolls in contact with the paper, depositing the ink. When printing a process color image, this happens four times, laying down successive layers of cyan, magenta, yellow, and black ink (CMYK).

As the first sheets of paper come off the press, the pressman inspects them, and they are compared with the proof you've supplied. The pressman has control over the amount of ink used in different areas of the page, but he cannot improve upon the image that was supplied to him.

Correcting Mistakes

Correcting a mistake or changing your mind about a picture you've chosen is always possible but can cost you dearly, especially if it's late in the process. To correct a scan after it's integrated into a page or presentation will only cost you the time it takes to readjust the scan or to scan the picture again. If you decide that the image needs adjustment or replacement after a digital proof is made, it will cost

an additional $25 to $75 for a new print. Changing the image after the film-based proof is made means paying for new film and a new proof, or approximately $200. Once the film goes to press, making a change can cost thousands of dollars. At this point most people learn to live with their mistakes. Table 2.1 gives you an idea of the costs that are associated with replacing photos.

Table 2.1   **The Cost of Replacing a Photo**

Production Phase	Estimated Expense
After Scanning	30 minutes of your time
After Digital Proof	$25–$75
After Film-Based Proof	$150–$250
On the Press	Thousands of dollars

Conclusion

As you can see, scanning is only one step in the complex process of reproducing an original transparency, photographic print, or piece of art. Just as a chain is only as strong as its weakest link, the final printed piece will only be as good as the care given in each phase of the production process.

Above all else, remember the principal of Starting at the End. Scanning, and the entire production process, must be done from beginning to end in the context of how the image will be reproduced. You can't start out in the right direction if you don't know where you're going.

CHAPTER

3

The Types of Scanners

Types of Scanners

Evaluating Your Needs

Software

Ascanner is nothing but a tool you use to do a specific job. Before buying a scanner, you need to evaluate your needs to make sure you get the right tool for the job at hand. Just as you wouldn't buy a tractor to plow a small home garden, or try to till a farm with a hand trowel, you don't want to buy too much scanner, or too little. Keep in mind that only a few of the many scanners on the market will actually meet your specific needs.

Choosing a desktop scanner can be a daunting task because there are literally dozens of models available, ranging in price from under $1,000 to over $30,000. As the price range indicates, not all scanners are created equal. But simply paying more won't ensure that you're getting the right scanner.

Types of Scanners

The seemingly infinite variety of desktop scanners on the market break down into three basic types;

- Flatbed scanners
- Transparency scanners
- Drum scanners

Flatbed Scanners

Flatbed scanners are the most common and least expensive scanners available. Not long ago, most models captured images as grayscale. Today, most flatbed scanners can capture images in color. Color models scan photgraphic prints and artwork by reflecting light off the surface of the print and directing light through filters to divide the image into its red, green, and blue (RGB) components. After passing through the filters, the light then hits a CCD (charged coupled device), which turns the light into electrical charges. Color flatbed scanners range in price from under $1,000 to over $10,000. Prices vary according to a scanner's resolution, quality, versatility, and the largest size original it can accept.

Flatbed scanners look like this:

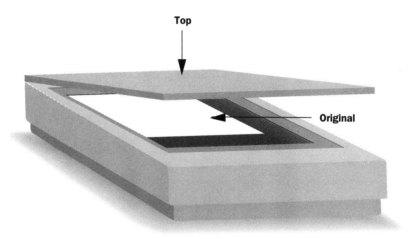

Transparency Scanners

Transparency scanners work similarly to flatbeds, except the light passes *through* the original; the original does not reflect the light. Prices range from under $2,000 for models that scan 35mm slides only, to $15,000 for those that can scan larger formats as well.

Below is an example of a transparency scanner:

Handheld Scanners

Several manufacturers make small scanners that you drag across the image you want to scan. These handheld devices are usually attached to an interface box, which in turn is attached to the computer.

Because of their small size and the relatively low quality of the images they capture, they don't really have any serious graphics uses. It can be fun, however, to scan the wallpaper or the pattern on your shirt.

While a few of these scanners are capable of scanning in color, they are mostly used for scanning text. Once the text is scanned in, it is converted from the pixels the scanner creates to editable text using OCR (optical character recognition) software. Once the "image" of the text has been processed through the OCR software, the text behaves as if it were entered from the keyboard. How faithfully the image is translated into editable text depends partially on the resolution of the scanner and the quality of the software. Most crucial, however, is the clarity of the original printed text.

Handheld scanners can be handy for scanning small pieces of grayscale artwork, such as logos. Scanning larger areas means tiling together several scans, which is not an optimal situation for either quality or speed. Several handheld scanners are designed to plug directly into laptop computers. This is convenient when you need to scan items that you can't take back to your home or office, such as public records or books in the reference section of a library.

The Transparency Option Feature

Many flatbed scanners on the market now offer a transparency option that lets you scan both transparencies and originals on the same scanner. Here's how this feature works.

Flatbed scanners reflect light off of the original. The original is placed underneath an opaque cover, which prevents the light from passing through it. The transparency option available for many flatbed scanners replaces the opaque top of the scanner with a light source. The light then shines from above through the transparency into the body of the scanner.

While the transparency option will do the job in a pinch, it usually doesn't yield very good results. Because transparencies can record more detail than prints, scanners made primarily for reflective art aren't designed to handle the tonal range of transparencies. Except in the higher-end models, you won't get the same quality scan as if you had scanned the transparency on a scanner designed specifically for scanning transparencies.

You might ask, **Can I use the tranparency option to scan 35 mm slides?** While you may get adequate results from scanning large size transparencies using the transparency option, it's nearly impossible to get a useful scan from a 35mm slide this way. Because 35mm slides have a much smaller image area than do typical reflective art, they require much greater enlargement. The greater the enlargement, the higher the scanning resolution needs to be. Flatbed scanners usually have a much lower maximum resolution than transparency scanners because they are designed for much larger originals. (See Chapter 4, "Resolution.")

If you need to scan both prints and slides, you could buy a desktop drum scanner (covered next) or a high-end flatbed with a transparency option, but either one will cost at least $15,000. A better solution is to buy two less expensive scanners, one just for prints, the other only for slides. Depending on your quality and speed requirements, you may be able to get both for under $4,000.

Desktop Drum Scanners

Desktop drums are a relatively new breed of scanners, which combine some of the quality of high-end drums with the affordability and convenience of desktop systems. Like their bigger siblings, desktop drums use PMT (photo-multiplier tubes) instead of CCDs for capturing the image. PMTs can sense a wider range of tones than CCDs, and therefore are capable of yielding better-quality scans. Prices range from around $15,000 to over $30,000. Desktop drums can accept both reflective and transmissive originals. (See "The Types of Originals," later in this chapter.)

Here's an example of a desktop drum:

You may wonder, **What is the difference between a desktop drum scanner costing $30,000 and a high-end drum scanner at a color trade shop, which may cost $150,000 or more?** There are really three things that account for the high price of high-end scanners. The first is simply the superior quality of the scanner itself. The high quality of the components and the greater care that goes into designing and constructing the scanner all contribute to making a superior product. The PMTs, the optics, and the mechanism that rotates the drum are all precision instruments. Their quality has a direct effect on the quality of the scans.

High-end scanners also cost more because they are built for high-volume production. Since they are designed for businesses that make their money directly from providing scanning and separations services, high throughput (speed) is a priority. Experienced scanner operators are well paid, so it makes sense to put them in front of the fastest machine possible. While you can buy ten desktop scanners for the price of one high-end drum, you'd need to hire ten scanner operators to run them. It makes more sense to put one scanner operator in front of a scanner that is ten times faster than a desktop scanner.

Lastly, the powerful workstation and software that comes with it contribute to the cost of high-end scanners. When you use a desktop drum, it's driven by software running on your personal computer. High-end drum scanners come with built-in computers running specialized software designed solely for driving the scanner and separating the image for process printing. Many high-end scanners even come with an integrated imagesetter for printing the film separations.

You should also be aware that many tasks that are accomplished through post-processing when a desktop scanner is used, are taken care of by the high-end scanner as the scan is being made. For example, such details as color and cast correction, sharpening, and separation are done "on the fly." This enables both faster scanning and higher-quality scans.

Evaluating Your Needs

As with most things in life, when it comes to scanners we want to have it all. We want to scan everything from 35mm slides to poster size artwork at the highest quality and as fast as possible. There are scanners that can do all this, but they are financially out of the reach of all but professional color shops. Most of us need to take an objective look at our needs and our budgets before buying a scanner.

In order to evaluate your needs, you need to ask yourself:

- What type(s) of originals do I need to scan?

- What are the quality requirements for the type of work I do?

- How important is fast throughput (speed) in my production process?

The next few sections are designed to help you answer these questions.

The Types of Originals

Consider how the originals you need to scan fit into these categories:

- Transmissive or reflective

- Positive or negative

- Color, grayscale, or line art

The following sections cover details about each of these groupings.

Transmissive or Reflective

Transmissive means that we view the image primarily by the light that passes through it. This would include transparencies of all sizes (including 35mm slides) and all negatives.

We view *reflective* originals with light that reflects off the surface. Photographic prints are reflective originals, as are artwork drawn or printed on paper.

Positive or Negative

We all recognize negatives when we see them. In a negative, the colors and tones of the image are reversed from how the scene originally looked. *Negative film* is used in cameras when photographic prints are to be made because prints look best when they are made from negatives. *Positive film* is used when the film itself will be viewed directly, as in a slide show. When photographs are taken specifically to be scanned, they are usually shot on positive film because positives provide an objective visual reference.

Color, Grayscale, or Line Art

Color refers to important hue information that an image contains. This is a fancy way of saying that it is colorful. In the world of computer imaging, the term *grayscale* corresponds to black-and-white photography. Grayscale images contain a range of gray tones. *Line art* is truly only black and white, with no gray tones or color. For example, a pen and ink drawing is usually considered line art.

Some scanners can only scan images as grayscale or line art. While there used to be a significant price difference between these scanners and those that can scan in color, today the price difference is minimal. If you are sure that you'll never need to scan color originals, you can save a few hundred dollars by buying a grayscale scanner. In most cases, however, it's worth the extra money to get the flexibility of a scanner that can capture color.

Sizes

Size is the final factor you need to consider when working with originals. Transparencies typically come in 35mm, $2^{1}/_{4}$", 4" × 5", and occasionally 8" × 10". Some odd size formats also exist. Photographic prints typically come as 3" × 5", 5" × 7", and 8" × 10". Artwork and printed material may be any size.

While every original will be some combination of the various categories, some combinations are more common than others. For example, reflective line art is fairly common while line art transparencies are a rarity. Negative transparencies are fairly common (especially 35mm size) while negative prints are almost never seen.

Quality Requirements

There's a saying in the world of audio equipment that you should never buy an amplifier that is better than your speakers. This is because no matter how good the amplifier, the music will never sound better than the speakers can reproduce it. The same principle applies to reproducing color. Furthermore, buying a higher quality scanner than you really need wastes money because there may be no visible difference between it and a lower priced model.

When evaluating the quality you need from a scanner, once again, "Start at the End" because the target printer or display determines the result. Think about the kind of final product you will typically be producing. For example, a high-quality scanner will not make your images look better if you plan to print on newsprint. High line screen printing, however, will readily show the difference between a low- and high-quality scanner.

Evaluating Quality

We can examine the quality of a scanner by looking for the following:

- Color fidelity

- Highlight and shadow detail

- Smoothness of tones

- Tight registration

- Repeatability

- Sharp focus

The following sections discuss each of these criteria.

Color Fidelity   The colors a scanner captures from a particular photograph depend on a combination of its light source, filters, electronics, and software. All of these factors vary from scanner to scanner. Some may have good fidelity in parts of the color spectrum, but have problems in other areas. Other scanners may have better overall fidelity yet are weak in some areas.

Holding Highlight and Shadow Detail   Traditional photographic film is really quite remarkable in its ability to hold detail in the highlights and shadows. *Highlights* are the lightest portion of the image that is not totally white, and *shadows* are the darkest part of the image that is not totally black. In fact, the range of tones found in photographic film is impossible to capture with most desktop scanners. However, you don't need to worry about capturing all the detail found in film because no printing method can reproduce the range of tones captured on film, either. But you do want the ability to capture detail in the most important areas of each image.

Scanners have analog components that introduce a certain amount of *noise*, or speckles, into the image. The scanner can filter out the noise, but as a result, some detail is lost. Some scanners can capture more information than is needed, in order to compensate for the noise without loss of detail. The scanner's ability to capture detail in the highlights and shadows is dictated by the sensitivity of its CCD. See the plate called "Tonal Reproduction" in the color section of this book.

Note. Because transparencies have a greater density range than reflective art, scanners designed to handle transparencies should be able to capture a greater range of tones than scanners designed for reflective art.

Smoothness of tones   The transition between tones is related to the range of tones a scanner can capture. If an area of an image consists of a gradual transition from light to dark, the image should reproduce smoothly. A scanner that can pick up subtle changes in tone can faithfully reproduce a shaded area without creating distracting and unnatural looking jumps between tones.

Registration   A scanner captures red, green, and blue (RGB) image information separately and then merges it in software. If the three channels of information are not properly aligned, the image is out of registration. Poor registration results in blurry images with visible color bands around the edges. Registration doesn't need to be off by much for it to be noticeable. Registration problems are usually the result of the imprecision of the mechanism that moves either the original or the CCD.

Repeatability   Most light sources need some time to stabilize. When a light is turned on, it may vary in intensity for a period of time, ranging from a few minutes to half an hour, until it reaches its point of stability. With some scanners, the light source is illuminated as soon as the scanner is turned on. This allows the light source to stabilize before you begin scanning. Other scanners only turn on their light sources when a scan is made. These scanners run the risk of exposing the first few scans differently from those that are made later, yielding different results from the same original.

Focus Normally, focus should not be a problem with flatbed scanners because the surface on which the original is placed should always be the same distance from the CCD that records the light. However, it's still important to take a critical look at how well a flatbed scanner focuses, especially in lower priced models, because sharp focus requires a very precise distance.

Focus is primarily a concern with transparency scanners, especially those that accept different sized originals. There are various ways in which transparency scanners handle focus. They may automatically set the focus, allow you to set the focus numerically, or let you adjust it visually. Some scanners make more than one method available to you. Because proper focus is controlled by minute changes in the distance between the original and the CCD, the scanner needs to allow very precise and accurate control.

Focus is affected by all sorts of things, such as the type of mount used, how much the film buckles, and the amount of heat that the scanner generates. Evaluate all auto-focus systems carefully, especially if this is the only type of focusing the scanner allows. Some of these systems work better than others.

What about Resolution?

If you listen to scanner salespeople, you can easily come away with the impression that the best way to compare the quality of scanners is to compare their maximum resolution. In fact, a scanner's maximum resolution has little to do with the quality of the scans it produces. The resolution simply tells you the number of pixels the scanner creates from each linear inch of the original; resolution tells you nothing about the *quality* of the image. While in theory, scanning at a high resolution should allow a scanner to acquire more detail, in practice a scanner's maximum resolution simply tells you how much it can enlarge an original (see Chapter 4, "Resolution").

Judging Quality for Yourself

The best way to judge the quality of a scanner is to actually use it. Information from the manufacturer is usually of little help because it tends to emphasize resolution or speed, avoiding the issue of quality, which is hard to quantify. In reality, most of us don't get the opportunity, or have the time, to test several scanners. Your best bet for obtaining information about the quality of scanners is to read reviews and comparisons in reputable computer, graphics, and imaging magazines.

Speed and Productivity

It once went without saying that all desktop scanners were slow. Where a few years ago a typical desktop scanner would take 10 minutes or more to scan a 5-megabyte image, today even some inexpensive scanners can do the same job in less than a minute. Depending on the kind of work you do, scanning speed may either be of utmost importance or incidental. A designer who spends several days compositing images into a photo-illustration may not care that it takes an hour to scan a few images. At a newspaper, on the other hand, time is of the essence. A slow scanner would mean moving up the deadline for photos, making them less timely.

Depending on your particular work flow, instead of buying a more expensive scanner for its speed, you may actually get faster throughput by buying two less expensive ones and attaching them to separate workstations. This way, you effectively double the speed of each scanner. Factors such as your quality requirements, the added cost of the additional workstation (or time on one you already have), will help determine if this approach is practical for your situation.

(Chapter 3, The Types of Scanners, continues after the following color section.)

Identifying Areas of Detail

Before an image is scanned, it must be evaluated for the important areas of detail. Knowing which areas are the highlights, midtones, and shadows will allow you to preserve detail in the important areas of the image as it goes from original, to scan, to print.

In this photograph, most of the important detail is in the texture on the building, which falls mainly in the midtones and quarter tones. You can see the detail of highlights in the birds in the foreground; the shadow detail can be seen in the darkest areas of the trees and water.

Highlight Detail

The highlight is not necessarily the lightest part of an image, but rather the lightest area where you want to maintain detail. The highlights can't be totally white, because some dots are needed to reproduce detail.

1

In this picture, the lightest areas of the greenhouse are "blown out" to white. As a result, the detail in those areas has been lost.

2

In this picture, the highlight detail has been preserved by making the lightest part of the green-house a bit darker than totally white.

Shadow Detail

Just as the highlight is not necessarily the lightest part of an image, neither is the shadow always the darkest part. The shadow is the darkest area of the image in which you need to preserve detail. If the shadow area is too dark, it will "fill in" and appear as a solid black area when printed.

1

In this picture, the shadow area on the right side of the canyon lacks detail.

2

In this picture, more tonal steps were used in the shadow area to preserve the detail on the right side of the canyon.

Tonal Reproduction

Photographic originals are continuous tone images. In order to preserve the illusion of continuous tone when scanning an image, a sufficient number of tonal steps must be preserved. The more tonal steps, the less abrupt the changes between tones will be when reproducing shaded areas.

1

In this picture, the tonal range of the image has been limited to less than 100 steps for each of the red, green, and blue channels.

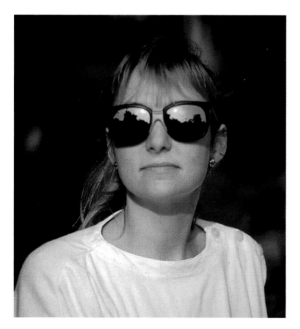

2

This image was reproduced using the full range of tonal steps. Notice how much smoother this image appears.

Color Cast

An overall color bias in an image is known as a *color cast*. Color casts can be hard to detect, especially if they're subtle. They're easiest to detect if the image contains certain memory colors, such as white, sky blue, green grass, and flesh tones.

1

This image has a greenish cast which was caused by the way the film was stored and processed.

2

This is the same image after its cast was neutralized.

Interpolation

If an image is scanned at too small a size or too low a resolution, it can be made larger through software interpolation, but at the cost of image quality. Interpolation works by making up new pixels based on existing pixels. When used in moderate amounts, the loss of image quality may not be too noticeable. When used to a greater degree, however, the loss of image quality becomes readily apparent.

1
This picture was scanned in at the proper resolution, so interpolation was not used.

2
This image was scanned in at only 75% of the correct resolution and then made one-third larger through interpolation.

3

This image was scanned in at only 50% of the correct resolution and then made twice as large through interpolation.

4

This image was scanned in at only 25% of the correct resolution and then made four times larger through interpolation.

Color Depth

The color depth of an image reflects the number of bits, or switches, used to control each pixel. The more bits used, the greater the potential number of colors or shades of gray for each pixel

I

In a 1-bit image, each pixel can only be either black or white.

2

In an 8-bit grayscale image, each pixel can be any one of 256 shades of gray.

3

In a 24-bit color image, each pixel can be any one of 16.7 million colors.

PhotoCD

Kodak's PhotoCD system uses a proprietary color model and compression scheme for storing images. The quality of the images when printed depends in large part on how they are acquired from the PhotoCD.

1

This image was acquired from a PhotoCD without using KCMS (Kodak Color Management System). It was then separated in Photoshop. Acquiring the image without KCMS results in the loss of some color fidelity.

2

This picture was acquired directly from the disk as a CMYK file using KCMS and Kodak's Precision Device Color Profile Starter Pack. This allows the image to more closely match the original slide.

Color Models/RGB

Film, scanners, and monitors all use the RGB color model. Red, green, and blue are known as the *additive colors* because adding them together creates white.

Composite RGB

Red

Green

Blue

Color Models/CMYK

Printing uses the CMYK color model. Cyan, magenta, yellow, and black are known as the *subtractive colors* because they create color by filtering out some of the light reflecting off white paper.

Composite CMYK

Cyan

Magenta

Yellow

Black

JPEG Comparison

JPEG is a method used for compressing images to greatly reduce their file size. As you can see from these examples, when JPEG is used the loss of image quality in the printed pictures is often negligible, except at the lowest quality settings.

1

Compression:
None
 File Size: 2.4MB

2

Compression:
High Quality
 File Size: 505K

3

Compression:
Medium
  File Size: 315K

4

Compression:
Low Quality
  File Size: 235K

Levels of Sharpening

Sharpening images is necessary to restore detail lost in the scanning process. Unsharp masking allows you to control the amount of sharpening and the areas that get sharpened. You want to sharpen areas with detail to enhance the edges, but not flat areas, because that would result in graininess.

1

This image has no sharpening. The entire image appears flat.

2

This image was sharpened without using any threshold setting, so the entire image was sharpened without regard to the amount of detail or flatness of an area. Notice how the sky and water have become grainy.

3

This image has the right amount of sharpening. Notice that the areas of detail, such as the rocks in the foreground, appear crisper, while the sky and water areas remain unchanged.

4

This image was sharpened too much and appears false-looking.

Moiré Patterns

Halftoning, the process of breaking an image down into discrete dots, is used for reproducing pictures in offset printing. These dots must be arranged in such a way that the image will appear to be continuous tone. If the dots are not arranged properly, the result will be a moiré pattern.

1

This image has a moiré pattern due to improper screening. Notice the uneven pattern on the sidewalk.

2

This image has proper screening and therefore, no moiré pattern.

Software and Prescan Controls

When considering the speed of a scanner, not only is it important to consider how fast the scanner captures the image, but you should also think about how well the scanner captures the image. This may sound like a quality issue but it has direct effect on productivity. Scanning a dozen photos in less than 10 minutes may seem fast, but if it then takes 15 minutes to color correct and retouch each image, the entire process is really rather slow.

Getting a good image that doesn't need a lot of color correction and retouching, also known as *post-processing*, is partially the result of having a good quality scanner—one that maintains color fidelity, can capture a broad range of tones, and is in focus. But getting scans that require little post-processing is also made possible by the software that drives the scanner. If your scanner is capable of processing more than 8 bits, the more image adjustments a scanner's software lets you make before you scan, the less post-processing the image will need, as long as the software can access the full bit-depth of the scanner. (More on scanner software later in this chapter.) Some software can also calibrate the scanner to the output device, giving you a scan that will need little adjustment. A few will even automatically separate your image into the process printing colors, which is a necessary step if the image will be printed on an offset press.

Some software allows you to gang scan images. *Gang-scanning* involves scanning several individual originals as if they were one large original, while giving you separate image files. Ideally, when gang-scanning, the software will allow you to adjust the settings for each original individually.

Removable Drums

For a desktop drum scanner, one of the most important features for enhancing speed is a removable drum. Often, the most labor intensive and time-consuming part of drum scanning is mounting the originals on the drum. By having a removable drum, originals can be mounted on a second drum while photos on the first drum are being scanned. As a result, the scanner never needs to be idle while images are being mounted. If you plan on getting a scanner with a removable drum to take advantage of the faster throughput, be sure to budget for the second drum, which can add several hundred dollars to the cost.

Software

Often when evaluating scanners, too much emphasis is put on the quality of the scanner itself, and not enough on the software that controls it. The software has a direct effect on the quality and productivity provided by the scanner.

Scanning Software

A scanner without software to run it wouldn't do much except take up space. Virtually all scanners come with at least the basic software you need for scanning. All color scanning software provides at least these basic functions:

- Prescanning
- Cropping
- Some type of control over resolution
- The ability to choose the original type as color, grayscale, or line art
- Brightness and contrast controls.

Other features that may be found in the scanner software include the following:

- Color balance

- Selection of the shadow and highlight points (either automatically or manually)

- Sharpening

- Setting the output size and resolution

- Negative to positive conversion

Additional features to look for are color previews and the ability to zoom in on previews. You should also note that software for transparency scanners will allow some sort of control over focus. As mentioned above, some software may even allow you to calibrate your scanner to your output device or automatically convert the scans to the process printing colors.

Here's an example of a software dialog box for version 4.5.1 of the Nikon LS 3510 transparency scanner for 35mm film:

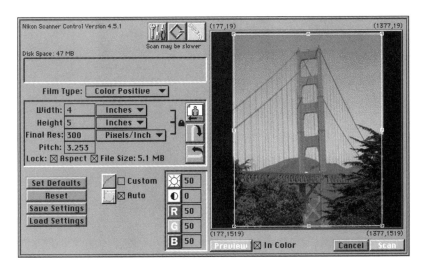

The software usually comes in at least two of the following forms:

- Stand-alone applications

- Plug-in modules for image editing applications, such as Photoshop for the Mac or Windows, and Picture Publisher for Windows

- The new standard called TWAIN (Technology Without An Interesting Name)

The TWAIN standard allows any compatible applications to access the scanning software. TWAIN is currently more popular with Windows systems than it is on the Macintosh.

Don't underestimate the importance of the scanning software. It has a direct effect on your ability to control the quality of your images, and the amount of time it takes to scan and correct them. The software is also the interface between you and the scanner, so it should be both functional and intuitive.

Bundled Software

Often, scanner manufacturers will include image editing software as part of a bundle when you buy a scanner. Again, the most common on the Mac side is Photoshop, and for Windows it is either Photoshop or PhotoStyler. This software can cost from $500 to $700 if purchased separately. If you need this software, getting it bundled with your scanner can be a great deal. However, if you already have the software or don't need it, it can unnecessarily add to the cost of the scanner. Some scanners have one price if they are bundled with image editing software, and a lower price without the software. To make matters even more confusing, there are also both full and limited versions of these applications, which come bundled with scanners.

Third-Party Software

Aside from the software that comes with a scanner, there are several third-party applications available that can drive scanners. These can

drive many types of scanners, and provide functions not normally found in bundled software, such as calibration, scripting, the ability to save settings, and even straightening crooked originals.

Conclusion

Which scanner is the best? There is no reasonably priced scanner that is perfect for all situations. You can find the best scanner for your needs by examining the types of originals you need to scan, how the images will be used, and how important speed is in your work environment. The right scanner is the one that meets your specific needs. Be careful not to buy a scanner based on what you think your needs might be in the future. You'll run the danger of paying for more scanner than you need, or worse, of getting one that is less than optimal for what you need today.

C H A P T E R

Resolution

The Anatomy of a Pixel

The Types of Resolution

Setting Resolution

Resolution and File Size

Real versus Interpolated Resolution

The Right Resolution

Resolution is one of the fundamental issues of scanning, yet it is also one of the least understood. Unless an image is scanned at the proper resolution, the results will be disappointing. Scanning at too low a resolution results in a coarse and pixilated image. Scanning at too high a resolution creates a larger file than necessary. This wastes time by taking longer to scan and print, and wastes money because it costs more to store a large image.

The subject of resolution can seem confusing, but in fact it's very straightforward. Unlike the issues of color and detail, which are subjective, resolution is quantifiable and therefore easy to calculate. But in order to do this, you need to be familiar with a few basic concepts. Taking the time to learn about resolution will pay off many times over in savings of both time and money.

The Anatomy of a Pixel

Photographic originals whether prints or transparencies, are continuous tone images. This means that the transition between tones and shapes is smooth and seamless. Unfortunately, computers can't directly understand the seamless variability of a continuous tone image because they can only process discrete bits of information. But a computer can work with an image if it's made up of dots. Scanning is the process by which we turn the continuous tone image into dots. These dots which we create when we scan are more properly called *pixels* (short for Picture Elements).

Pixels are the building blocks of digital images, and you need to consider the following factors when working with them:

- Size

- Location

- Value

- Bit-depth

The following sections cover each of these issues.

Size

A pixel's size determines the amount of information in a scan. The size of a pixel is determined by the resolution at which the image was scanned. The smaller the pixels, the more detail the image contains.

Location

The pixels that make up an image are arranged on a grid with a horizontal and vertical axis. Each pixel has a specific, measurable location on the grid. For instance, a pixel's location may be described as 250 pixels from the left and 300 pixels from the top. The number of cells in the grid, and therefore the number of pixels that make up the image, is determined both by the image's resolution and its physical size. The higher the resolution, the more pixels in a given area. The larger the physical size of the image, the greater the area the grid covers.

Value

When a photograph is scanned, each pixel is given a value based on the color of the point on the original that it represents. All color scanners acquire color information as varying combinations of the additive colors, red, green and blue (RGB). These colors can be combined to represent virtually the entire visible spectrum. Each pixel can only be one color, but when many small pixels of progressively different values are placed next to each other, it creates the illusion of continuous tone.

Bit-Depth

The last characteristic of a pixel is its bit-depth. A pixel can only have one value. The bit-depth determines the number of possibilities for the pixel's value. The greater the bit-depth of the pixels, the more data required to store the image. (This subject is covered at length in Chapter 6, "Understanding and Using Color.") The important thing to keep in mind for now is that the greater the pixel's bit-depth, the larger the file size will be.

The Types of Resolution

When we talk about resolution in terms of digital imaging, we need to consider the following types:

- Scanning resolution

- Output resolution

- Printer resolution

Not understanding the difference between these three types of resolution is probably the most common reason people get confused when dealing with resolution. Let's look at each one individually.

Scanning Resolution

As we have already noted, scanning turns a continuous tone image into pixels. The scanning resolution is the number of pixels created from a specific area of the original. For example, scanning an original at 300 ppi (pixels per inch) will yield 90,000 pixels (300 pixels horizontally × 300 pixels vertically) from each square inch of the original.

Adjusting the scanning resolution allows you to control the number of pixels created and therefore the amount of information captured.

Chapter 4: Resolution

Calculating Scanning Resolution

The total number of pixels created from a scanned image is determined by both the scanner's resolution and the physical size of the original. Scanning a 4" × 5" print at 100 ppi will result in 200,000 pixels. Scanning an 8" × 10" print at the same resolution will yield 800,000 pixels. Therefore, the relationship between the scanning resolution and the output resolution is based on the size of the original. To reproduce an 8" × 10" original with an output size of 8" × 10" and an output resolution of 300 dpi (dots per inch), it only needs to be scanned at 300 ppi. To reproduce a 4" × 5" print with the same output size of 8" × 10" and output resolution of 300 dpi, it needs to be scanned at 600 ppi. The smaller the original, the higher the scanning resolution needs to be.

Table 4.1 provides an example of how you can calculate scanning resolution.

Table 4.1   **How to Calculate Scanning Resolution**

Size of Original	Output Size	Output Resolution	Scanning Resolution Required
1" × 1.5" (35mm slide)	8" × 10"	300 dpi	2,190 ppi
4" × 5"	8" × 10"	300 dpi	600 ppi
8" × 10"	8" × 10"	300 dpi	300 ppi

Given this basic information, you might ask, **Why does the resolution of typical flatbed scanners range from 300 to 1,200 ppi, while even the least expensive slide scanners can scan at close to 2,000 ppi?** The answer lies in the fact that transparency scanners need to have higher maximum resolution than flatbed scanners because transparencies are usually smaller than reflective art. A 5" × 7" original reproduced at an output size of 5" × 7" inches to a printer that requires an output resolution of 400 dpi needs to be scanned at only 400 ppi. To reproduce a 35mm slide at 5" × 7" to the same device requires that the slide be scanned at almost 2,000 ppi.

The resolution of multiformat transparency scanners is often described not in dots per inch, but in the total number of pixels the scanner can capture both horizontally and vertically. At maximum resolution, a scanner may acquire 4,000 by 6,000 pixels, regardless of the size of the original. A 35mm slide (approximately 1" × 1½") when scanned at maximum resolution would result in 4,000 pixels captured per inch of original, while a 4" × 5" transparency would yield only 1,000 pixels from each inch of the original.

Here's another common question you may ask: **When comparing two similar scanners, will the one with a higher maximum resolution give me better quality scan?** The biggest myth of scanning is that higher resolution means higher quality. In fact, resolution has less to do with quality than it does with enlargement. The higher the resolution of a scanner, the greater it can enlarge an image from its original size to the output size.

While scanning an original at a high resolution will theoretically allow you to capture more detail, the amount of detail that can be reproduced is determined by the resolution of the output device. Scanning at too low a resolution will produce visibly poor results, but scanning at too high a resolution will take longer and cost more to store and print, while looking no better than if it were scanned at the correct resolution.

The higher-resolution scanner does allow you to enlarge the original to a greater degree than the lower-resolution scanner. By allowing more enlargement, this scanner also lets you scan only small parts of an original and blow it up to a useful size.

By itself, a high maximum resolution doesn't ensure high quality. (The only exception to this rule is line art. For more on scanning line art see Appendix D, "Line Art".) When deciding what the maximum resolution of your scanner needs to be, consider the size of the originals you plan to use, the amount you plan to enlarge (or reduce) them, and the output resolution required by your printer.

Output Resolution

The output resolution is the resolution at which the image file needs to be sent to a particular output device. This can be confusing because the output resolution is not necessarily the same as the resolution of the printer. The output resolution is determined both by the resolution of the printer and whether the device uses a halftoning method or reproduces images as continuous tone.

Printer Resolution

Printer resolution is the maximum number of dots per inch the printer can create. For instance, a 300 dpi (dots per inch) laser printer can create dots that are $1/300$ of an inch. A 2,500 dpi imagesetter can make dots as small as $1/2,500$ of an inch.

All printers can be characterized by the number of dots per inch they are capable of reproducing. How the printer resolution relates to the output resolution depends on whether the printer uses a halftoning or a continuous tone technique for reproducing images.

Halftone Printing

If you want to know the correct output resolution when going to a halftoning device, the short answer is $1^1/_2$ to 2 times the line screen. If you want to know why this is the case, read the next two paragraphs.

When calculating the output resolution for a scan that will be printed on a halftoning device, the output resolution is based upon the line screen, not the printer's resolution. There is considerable debate in professional prepress circles as to what the relationship should be between output resolution and line screen. The debate consists of whether the output resolution should be double the line screen, or only $1\frac{1}{2}$ times the line screen. Using the example of a 150-line screen print job, some experts say that the output resolution should be 300 ppi, while others say that 225 ppi is sufficient. It pays to test both of these ratios out for yourself. If you can't see a difference in the final printed piece, go with the ratio of $1\frac{1}{2}$ times the line screen. It will take less time to scan and produce a smaller file. If you can see a difference, use the higher ratio of 2 times the line screen.

If the ratio were only 1 to 1, you'd run the danger of ending up with a pixel from the scan not aligning properly with the halftone screen. The result would be that some pixels acquired from the scan wouldn't be represented by halftone dots, while other pixels would be represented by more than one halftone dot. Therefore, the outcome would be a rough looking image because some pixels would have been lost and others would have been duplicated.

For more details on halftoning, see Appendix A.

Continuous Tone Printing

Some color digital printers can create output that is so smooth it appears to be continuous tone. Dye-diffusion, some color laser printers, and high-end inkjet printers use various methods to apply the ink, dye, or toner in such a way that they are blended either before or at the time they hit the paper. Unlike halftoning devices, which create shades, colors, and shapes by carefully adjusting the size of discrete dots, continuous tone devices reproduce images with pigments that blend together and overlap.

Continuous tone devices don't require the images to be broken down into a line screen. Therefore, the output resolution that these devices require is more directly related to the printing resolution than is the case with halftoning devices. In theory, the output resolution should be the same as the the printer's resolution. For example, when printing to a 400 dpi color laser printer, the proper output resolution should be 400 ppi. In actual fact, it's difficult to see the difference between an image sent to that printer at 400 ppi and one sent at 300 ppi. Below 300 ppi, there won't be enough data to simulate continuous tone, resulting in a choppy looking image.

Each type of continuous tone output device has its own output resolution requirements. While it's generally safe to send an image at 75% of the printer's resolution, your best bet is to ask the manufacturer of your printer for the correct output resolution, or to ask the service bureau or prepress facility for the resolution required when using their printer.

Setting Resolution

Some software automatically sets the scanning resolution for you. You simply enter the size you wish the printed image to be, and the output resolution required by the printer. For example, let's assume we need to scan a 4" × 5" print that will be reproduced at 8" × 10" to a device that requires an output resolution of 300 ppi. You would simply enter your output size as 8" × 10" and your output resolution as 300 ppi. The software will automatically set the correct scanner resolution, in this case 600 ppi.

Other software won't automatically calculate the proper scanning resolution for you; you will need to choose the scanning resolution directly. In this case, you must do your own calculating to arrive at 600 ppi for the scanning resolution.

Figure 4.1 shows a dialog box from version 4.5.1 of the software for the Nikon LS 3510 transparency scanner (35mm film, only).

Resolution and File Size

Every digital image is made up of specific number of pixels, and the way the pixels are distributed determines the image's physical size and resolution. If the same number of pixels is spread over a large

Figure 4.1
A scanning
software
dialog box

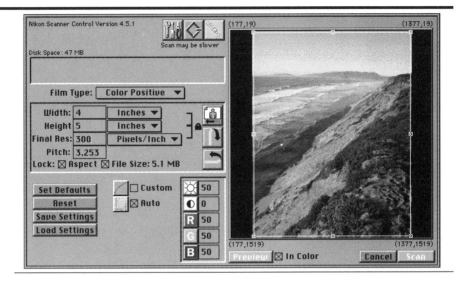

area, the image is large, but has a low resolution. If the pixels are concentrated into a smaller area, the image is small but has a higher resolution.

The amount of data needed to store an image, or its file size, is determined by the number of pixels that make up the image. A large image at a low resolution may be the same file size as a smaller image at higher resolution. For example, an RGB image that is 700 × 1,000 pixels takes up 2MB of disk space. This image can be reproduced at 7" × 10" at 100 ppi, or 3½" × 5" at 200 ppi, just to name a few options.

Real versus Interpolated Resolution

The resolution of a scanner is often rated in two different ways: by its real—or optical—resolution and by its interpolated resolution. The optical resolution is based on the number of elements in the scanner's CCD (charged coupled device), the part of the scanner

that turns light into the electrical charges, which eventually become pixels. Interpolated resolution is based on the number of pixels that are created when the scanner or its software artificially multiplies pixels during scanning.

Interpolated resolution is typically double or quadruple the optical resolution of the scanner. For instance, a scanner that has a true optical resolution of 400 ppi may have an interpolated resolution of 800 or even 1,600 ppi. An original scanned on such a scanner at higher than 400 ppi will not have the same quality as if were scanned at 400 ppi or lower. A scanner and its software varies greatly as to how much the image suffers by using interpolated resolution. With some scanners, it is hard to tell that interpolation was used, and you get the benefit of capturing more detail and greater enlargement. With others, the loss of quality is readily apparent when scanning at higher than the optical resolution.

An image's resolution can also be interpolated with most image editing software. Programs like Adobe Photoshop and Aldus PhotoStyler let you increase the size or resolution of an image by creating new pixels. But these pixels created through interpolation are not based on the original image, as they would be if they were created during scanning. Instead, the new pixels are based upon the pixels that already exist in the image. These programs use very sophisticated methods to calculate how these new pixels should look. But it's still just guesswork. An image that is made larger through interpolation will not look as crisp, or be as faithful to the original, as if it were scanned at the proper resolution to begin with.

When images are greatly enlarged through interpolation, they break down and become fuzzy or pixilated. (See the plates called "Interpolation" in the color section of this book.) Software interpolation should be avoided or used sparingly. The best way to avoid having to use interpolation is to acquire enough pixels at the time you scan, by scanning at the correct resolution.

The Right Resolution

When selecting any type of resolution, there are really only three options: too low, too high, and just right. The important thing to remember about resolution is that despite all the choices, each scanning situation has only one correct scanning resolution and one correct output resolution.

You may wonder: **How do I decide what the scanning resolution should be if I don't know the size the image will be when it's printed, or what device will be used to print it?** The answer is that the right scanning resolution is based on the image's final size and the resolution required by the printer. While it's always better to know these things before you scan, sometimes it's just not possible. In most of these cases, your best bet is to scan at the highest resolution that is practical given the amount of time you have to spend scanning and the disk space you have to store the image. Just as software interpolation can be used to enlarge images, it can also be used to reduce them. Reducing the size or resolution in software does not degrade the quality of the image nearly as much as enlargement. Still, it does affect the quality of the image because shrinking it down will cause it to lose contrast. As always, scanning at the correct resolution, when possible, will yield the best results.

In some situations, scanning at too high a resolution is not practical because of the time it takes and the large files it creates. For example, when newspaper photographers scan a photograph and transmit it back to the office, they rarely know at what size the image will be used. But they also can't waste time by scanning at a higher resolution than necessary, or by having to transmit large files via modem.

A good strategy to use in this situation is the 80/20 rule. This entails scanning images at a resolution that meets or exceeds your

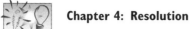

needs at least 80% of the time. This way, the scans you make will meet most of your needs without being unnecessarily large. Even if you need to rescan 20% of the images at higher resolution, you'll end up saving more time than if all the images had been scanned at the highest resolution you may potentially need.

Conclusion

The whole issue of resolution is more complex in theory than it is in practice. If you know the output required by your line screen or printer's resolution, and you know the final physical size of the image after it is reproduced, you will have everything you need to scan at the correct resolution.

CHAPTER

5

Step-by-Step Scanning

Starting Smart
Before You Scan
Post-Processing

Desktop scanning, like any production process, is made up of a series of interrelated steps. How each one is carried out has an impact on the final printed image. By carefully and methodically going through the steps of scanning, you'll be able to get consistent, predictable results.

Starting Smart

Before we discuss the actual scanning process, let's first look at a few important points that are sometimes overlooked. Neglecting these fundamentals will undermine all the care you put into actually scanning the image.

Placing the Scanner

You may not have given a lot of thought about where your scanner should be located. This is a mistake because the placement of your scanner is critical to both the quality of the scans and the health of your scanner. When setting up your scanner, keep the following details in mind:

- *Place the scanner on a stable, level table or counter.* Scanners are precision instruments with parts that move in steps smaller than a thousandth of an inch. Even slight movement or vibration will jar the moving parts of the scanner, causing visible flaws to appear in the scanned image. Excessive movement or vibration can actually cause permanent damage to the scanner by forcing the delicate parts out of alignment.

- *Be sure that the air vents are kept open and clear.* Heat is the enemy of all microcircuit equipment. Most scanners produce a significant amount of heat that is generated by their bright light source. Excessive heat can cause permanent damage to your scanner and may even result in immediate failure. At the very least, too much heat will shorten the scanner's life. A hot scanner also introduces "noise" into the darker areas of the image, resulting in a loss of shadow detail. If the scanner gets too hot, it can even damage the original, causing it to fade and possibly even blister or melt. To avoid these kinds of problems many scanners (especially transparency scanners) have fans and air vents. It is essential that you leave sufficient clearance around these vents because they allow air in to help cool down the scanner.

- *Leave plenty of room for the scanner to do its thing.* Many scanners have moving parts outside the scanner's main body. Several flatbed scanners in particular have moving beds that may require up to twice as much space as that taken up by their footprints. It's also a good idea to have extra desk space nearby for keeping the originals to be scanned, cleaning supplies, and a light table or viewing booth. The shorter the distance you need to move the original after inspection and cleaning, the less chance it will get dusty or damaged before the scan.

- *Work in a dust-free environment.* Originals will invariably have some dust on them, but you should protect them as much as possible. Besides, dust isn't very good for your scanner either. In a dusty environment, dust particles can get sucked into the scanner, resulting in spots on your scans. Once dust gets inside a scanner it's difficult to remove; over time, too much dust in your scanner can cause permanent damage.

Tools for Viewing and Inspecting Originals

Originals need to be inspected before they are scanned. For inspecting transparencies, you'll need a good light table and loupe. The light table should be clean and smooth. Don't use the same light table that's been used for doing paste-up work or stripping. Chances are, it will be full of scratches from X-acto knives and goo left over from rubber cement and lithographic tape. For reflective art, you should use a color viewing booth.

The light table and viewing booth should also be evenly lit, preferably with a light source designed for viewing color. Because color is made up of light, you should make color decisions using *balanced light*, which is light made up of equal amounts of all visible colors. The color of a light source is described in degrees on the Kelvin scale. A neutral light source with no color bias is designated 5,000 kelvin. Sometimes other types of light are used for viewing color. Whatever type of light is used, it's important to use the same type of light when inspecting the original, evaluating the proof, and checking the press sheets. (For more information on viewing color, see Chapter 8, "Calibration.")

A loupe is also essential, especially if you'll be scanning 35mm slides or negatives. When you scan an original of that size, you will likely be magnifying it many times. Inspecting it through a loupe allows you to see any dust or flaws that would escape detection with the naked eye.

Before You Scan

There are still several important things that need to be done before you're ready to scan. Once an image is selected to be scanned, it must be evaluated from a production standpoint.

Physical Inspection

Every original needs to be inspected carefully for dust and physical imperfections, such as scratches and fingerprints. Unfortunately, most originals will already have some physical flaws by the time they are ready to be scanned. When possible, dust should be removed and flaws should be corrected before scanning. Anything on the original will be picked up by the scanner and will usually be magnified.

Dust

Even if an original has been handled with the utmost care, it will still likely have at least some dust on it. Therefore, removing dust should be a regular part of every scanning process. Fortunately, this is a fairly straightforward procedure. Simply blow the dust away using either a blower brush (a brush with a rubber bulb at the end that blows air when you squeeze it) or canned air. If you use a blower brush, make sure it is of good quality. The bristles of a cheaper brush can scratch the film.

Canned air is usually a better solution. Canned air is not really compressed air, but rather a chemical propellent that expands and creates a blowing action when it is released. When using canned air, always keep the can in an upright position. Tilting the can during use will cause the propellent to come out as a liquid, damaging the original.

CAUTION! *Don't ever just blow on an original to remove the dust. Your breath contains moisture that can damage the original.*

Scratches and Fingerprints

Photographic prints or other reflective art can be scratched if they are badly handled. But scratches are much more likely to be found on slides, transparencies, and negatives. A small scratch on an original may only be a minor inconvenience. However, a large scratch can be a major problem, making the original unusable. How you deal with scratches and fingerprints depends on whether they are found on the emulsion or non-emulsion side of the film.

Emulsion versus Non-Emulsion   Film consists of two primary parts, a light-sensitive layer known as the *emulsion* and an acetate backing; color film actually contains several layers of emulsion. The emulsion is where the image information is actually recorded. The backing is simply there to hold the emulsion. The emulsion is less durable than the backing and therefore more susceptible to scratches and absorbing oil from fingerprints.

The emulsion is often identified as the dull side of the film because it is less shiny than the backing. In fact both sides appear fairly shiny so you usually need to compare them to find the emulsion side.

Removing Scratches   You can use any image editing software to easily repair small, unobtrusive scratches in a flat (low-detail) area of an image after scanning. Large scratches, or those that go through important high-detail areas, are a different story.

If the scratch is on the backing side, you can repair it by carefully filling it in with a small amount of Vaseline. This is a delicate procedure that requires some experience to master but it's very effective. Whatever you do, don't get any Vaseline on the emulsion side, because irreparable damage will result.

If the scratch is on the emulsion side of the film and goes through an important area that contains a lot of detail, try to find a different original to take its place. If that original must be used, plan on spending a lot of time retouching it with an image editing program. Even then, the results are not guaranteed. You must weigh the importance of using that original against the hours it may take to digitally remove the scratches.

Handle with Care

Professional color houses go to great lengths to protect originals and so should you. Because you sometimes need to go to great lengths to clean and repair damaged originals, you obviously do not want to add dirt or otherwise damage them.

Always handle every original by its edges to avoid adding scratches or fingerprints of your own. It's a good idea to always wear lint-free cotton gloves when handling film so you don't damage an original by accident.

Removing Fingerprints Fingerprints tend to be larger and more difficult to remove than most scratches. If the fingerprints are on the non-emulsion side of the film, you may be able to get rid of them by using a commercial film cleaner and lint-free tissue, both of which are available at most photography supply stores. Dampen the tissue with film cleaning solution and gently dab the film. Do this carefully to avoid scratching the film.

If you find a fingerprint on the emulsion side, you're out of luck. The oil from a fingerprint is readily absorbed into the film's emulsion and is thereafter impossible to remove. In addition, most fingerprints are too large and complex to remove easily with an image editing program. An original with fingerprints on the emulsion side is virtually unusable.

Evaluating the Content

Once the film is inspected for physical flaws and cleaned or repaired, the next step is to evaluate the content of the image. You first need to decide what's important in the image. Because some information must be lost or changed, it's important to determine which areas must be reproduced as faithfully as possible as well as those areas that can shift in color or lose detail. Print reproduction entails making a series of compromises in color and detail from the original scene until it's reproduced as final output. Rarely can you completely match printed output to an original.

Once you decide what's important in the image, it's time to evaluate the quality of the original. This evaluation is based on the following factors:

- Exposure

- Color cast

- Sharpness

These three issues are covered next.

Exposure

You need to recognize an image's highlights, shadows, and midtones before you can judge its exposure. The *highlight* is the lightest area of the image that contains detail. The *shadow* is the darkest area of the image that contains important detail. The *midtones* are the range of tones that fall between the highlights and shadows. See the plate "Identifying Areas of Detail" in the color section of this book.

High-Key, Normal, and Low-Key Photos   Some photographs are intentionally taken in a way so that all the important detail is exclusively in either the highlights or in the shadows. Images that are exposed so that the important detail information is in the highlights are known as *high-key* photos. *Low-key* images have most of the important detail in the shadow area. Normal photos are by far the most common types of images you'll come across. In *normal* images, the important detail is either contained in the midtones or it is evenly distributed throughout the image's tonal range.

The character of the detail in a photo should not be confused with the photo's exposure. A properly exposed original will capture all of the important detail in a scene. An underexposed image will be dark overall and lack detail in the shadow area. An overexposed image is light overall, lacking detail in the highlight area.

Remember, most types of output cannot display the wide range of tones that can be present in a photographic original. Through experience you will learn about the capabilities of your output device, which will make it easier for you to select originals accordingly.

Color Cast

A *color cast* is a uniform color bias of the original. It may be either very subtle of very pronounced. You can detect a color cast by examining the neutral tones in the image. The cast is a result of the particular combination of film, processing, and lighting used to take the picture.

Sharpness

Despite the fact that scanning software and image editing software come with all kinds of tools for sharpening, if an original is not in focus there's not a thing you can do to make it so. The sharpening tools available in software are only for restoring sharpness lost in the scanning process, not for bringing out-of-focus originals into focus.

In many photographs only some of the image is in focus in order to emphasize specific elements in the picture. In cases like these, be sure that the important areas of the image are sharp to avoid disappointing results.

Detecting Casts

Q: How do I tell if an original has a cast?

A: Color casts are most obvious in the neutral tones—the white and gray areas. The best way to detect and evaluate a cast is to view the original with a 5,000 kelvin light source. (For a description of this light source, see "Tools for Viewing and Inspecting Originals," earlier in this chapter.) If you can't see any tint in the neutral areas, try holding a gray card, available from any photographic supply store, next to the original. These cards are printed using only black ink, so any color in the neutral areas will become obvious when compared to the gray card.

A subtle color cast is easy to correct either during the scan, using controls in the scanning software, or after the scan, using image editing software. A severe cast can make it difficult, if not impossible, to reproduce the image without false-looking color. In either case, evaluating the cast before you place the image in the scanner is important because it allows you to detect the color and severity of the cast and plan how to compensate for it.

Not all casts require correction. Sometimes a neutral area takes on a cast from the color that surrounds it. For example, in a picture of a white ball in a red room, the ball will pick up a reddish cast. This is natural and reflects the way the scene would look if we were there to view it. We expect to see a white object take on some of the color that surrounds it. Removing the red cast from the image would make the white ball look false.

Now You're Ready to Scan

You're finally ready to begin scanning. A step-by-step scanning guide such as this needs to be somewhat generic because the capabilities and controls of various scanners and their software all differ. But no matter what specific combination of scanner and software you use, all of the following steps need to be done.

Placing the Original in the Scanner

Flatbed scanners all handle originals in a similar manner. The print is placed on the glass bed face down. The correct position (flush left, flush right, or centered) is usually indicated on the scanner. In any event, make sure you place the original as straight as possible because this takes much less time than correcting a crooked scan using image editing software.

Transparency scanners vary in the way they handle originals. Some allow you to put in strips of negatives, others let you put in mounted slides, some let you do both. Yet others require that you remove a slide from its mount.

Scanners that can accept multiple film formats usually require that you use special film holders when inserting the originals into the scanner. Many dedicated 35mm slide scanners allow you to simply insert the mounted slide as you would put bread into a toaster.

Originals are mounted on the drums of drum scanners using either oil, tape, or powder. When oil is used, the original transparency is dipped in the oil and then carefully placed on the drum. A piece of clear acetate is then placed over the transparency. Great care must be taken when handling the original because when the emulsion is soaked with oil it swells and becomes soft, making it prone to severe scratches. The procedure for using powder is similar to that for oil, but the use of powder reduces the risk of damaging the original.

After the original is placed on the drum, any air pockets must be carefully removed. Otherwise, spots will appear on the scan, or circular rainbow patterns, know as Newton Rings, will be visible.

Focus

Flatbed scanners generally don't require focusing, so they don't need to give you controls over focus. Transparency scanners and drum scanners may offer a variety of ways to adjust focus. The scanner may either set the focus automatically, allow you to set it manually, or give you a choice. However the scanner lets you set the focus, make sure that the focus is as sharp as the scanner will allow.(See Chapter 3, "Types of Scanners," for more on focusing scanners.)

Prescanning

Prescanning creates a low-resolution preview of the image. Scanning software differs in the amount of previewing flexibility.

Some programs let you do each of the following:

- See previews in color

- Change the size of the preview window

- Allow you to zoom in on the preview

Various software offers all or some of these options.

Cropping the Image

After prescanning, crop the image in the preview window. Cropping an image before it's scanned will make the scan go faster because time won't be wasted scanning unneeded areas of the original. This also speeds up post-processing by not forcing you to crop a large file after the scan.

Setting Resolution

Different scanning software gives you various options for setting the resolution of the scan. Some software lets you set the output size and output resolution, and will automatically calculate the correct scanning resolution. Others require that you figure out the correct scanning resolution and enter it directly. (For more on setting scanning resolution, see Chapter 4, "Resolution.")

Most scanning software allows you to lock the size and resolution settings (Figure 5.1). If you do this, you can change the crop and still be sure of getting the correct output size and resolution.

Adjusting Color and Contrast

Color and contrast adjustments can be made either with the scanning software before the scan is made, or with image editing software after the scan. Scanners that can recognize more than 8 bits of data from each RGB channel will yield much better results if color and contrast adjustments are made before the scan rather than afterward. Even with lower quality scanners, it's usually better to make these adjustment before you scan. If your scanning software doesn't allow you to make these adjustments, you'll need to do them afterwards with your image editing software.

Figure 5.1
Once you set the size and resolution of a scan, you can lock the settings. This allows you to change the crop without changing the settings.

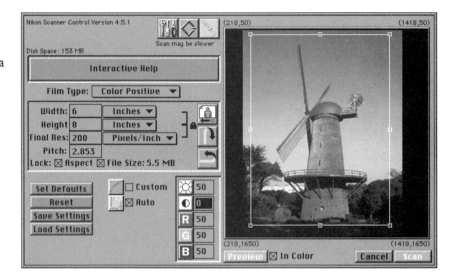

Setting Contrast Range   The best way to set the contrast is by selecting the image's highlight and shadow point. When you do this, you're telling the scanner what you consider to be the important highlight and shadow detail. For instance, selecting the lightest part of a white tee shirt tells the scanner that point is the lightest area in which detail needs to be preserved. Most scanners allow you to do this automatically by clicking on an Auto Exposure button. How well this works depends on how "typical" the original is that you are scanning. Avoid using auto exposure settings if the original has dust and scratches on it because these flaws can fool the software into reading false highlight and shadow areas. You may want to try the auto setting first and then use the manual selection if you're not satisfied with the results. Many scanners also let you manually select the highlight and shadow points by clicking on the areas in the image that you want to represent those points.

Adjusting Color   Scanning software usually enables you to individually control the amount of red, green, and blue (RGB) acquired from the scan. It's important to remember that the more color you remove, the darker the image will be. This is because in the RGB color space, the colors are added together to make white. Therefore, removing any color will make the image darker.

Be careful of making color decisions based solely on what you see on the monitor. Unless you've calibrated your monitor to your final output, it won't necessarily display colors accurately. Even if you have calibrated your monitor, it displays colors much differently than does printed output. The best way to make color decisions is to learn how to read color percentages using the densitometer in your scanning or image editing software. However, this takes a good deal of experience to master. Lacking that, use a properly calibrated monitor and color proofs for making color decisions. (For more on calibrating your monitor, see Chapter 8, "Calibration.")

Scanning

Once everything is set, acquiring the scan usually involves simply clicking on the Scan button in the scanning software. Make sure the scanner remains undisturbed during the scan. Don't lean on the table that the scanner is on and protect it from any strong vibrations.

Post-Processing

Now that you've acquired the scan, there are still a few steps that you need to perform. These steps involve using image editing software.

The First Save

Some scanning software requires that you save the scan directly to your hard drive, which saves it as soon as it's scanned. Other software, especially plug-in modules, will simply display the image on the screen once it's scanned, without saving it to disk. In this case, the first thing you should do is save the image. This is a good habit to develop because otherwise you'll have to rescan the image if your computer crashes or if you make any undesirable and irreversible changes to the image.

Final Crop

Often the preview window in scanning software is too small to allow you to precisely crop the image before scanning. If this is the case, the image should be given its final crop after it is scanned and saved for the first time, but before spotting for dust. An image should always be given its final size and resolution before it's imported into a page layout program. Otherwise, you risk having the image print at the wrong resolution.

Spotting for Dust and Correcting Scratches

Even if you had carefully cleaned the original before scanning, it will still likely have had some dust on it. This dust will leave spots on the scanned image. One of the most efficient ways to remove dust spots is to use the image editing program's Rubber Stamp tool. Simply select an area near the dust spot of similar tone and color, and copy that area onto the dust spot. This is easy to do for background and solid areas. It may take a bit of practice to master retouching high-detail areas. The same method can be used for correcting any scratches you find on the film.

Final Contrast and Color Correction

If you were not able to make sufficient contrast and color correction in your scanning software, do it now in the image editing software. Most image editing software offers a variety of ways to control contrast and adjust color. Once again, don't rely on an uncalibrated monitor for making these corrections. Even if your monitor is calibrated to your output, it's better to rely on proofs.

Figure 5.2 shows some dialog boxes in Photoshop that are used to correct color and contrast.

Unsharp Masking

Unsharp masking accentuates the edges of objects in an image by selectively increasing the contrast between neighboring pixels. This technique is necessary when reproducing any image in print. Unsharp masking differs from sharpening in that sharpening accentuates the contrast between *all* the pixels. While this produces the desired sharpening effect in the areas of detail, it also adds graininess to smooth areas, such as backgrounds and flesh tones. Unsharp masking lets you control where this increase in contrast happens.

Figure 5.2
Photoshop,
like most
image editing
programs,
gives you
several ways
to adjust the
contrast and
color of a
scanned
image.

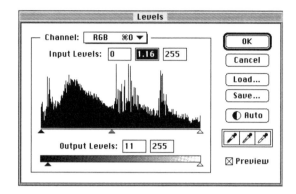

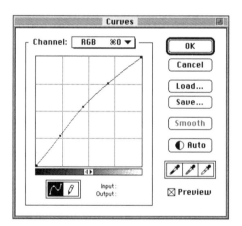

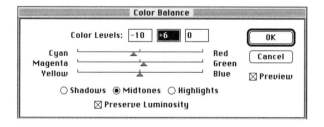

Unsharp masking is usually done with image editing or separation software, although some scanning software can do it as well. Unsharp masking settings let you control the amount and threshold of sharpening and the threshold of pixels. The threshold designates the point at which neighboring pixels have different enough values to be considered an edge, and therefore have their contrast accentuated. Software with more sophisticated controls allow you to set a level of softening, a slight decrease in contrast, for areas that fall below the threshold. See the plate "Levels of Sharpening" in the color portion of this book.

Why "Unsharp"?

Q: Why is it called "unsharp" masking if it actually makes the image look sharper?

A: The term unsharp masking is a holdover from the days of traditional halftoning. Before halftoning was done digitally, the camera operator would shoot the image once, and then shoot it again slightly out of focus. When the two images were merged, the difference between the sharp edges of the in-focus film and the blurry edges of the "unsharp mask" provided the added contrast needed to accentuate the edges in the image.

Separation

If your image will be printed on an offset press or to a digital color printer that requires that a CMYK file be sent to it, you'll need to separate the image into the process printing colors. You can either do this in your image editing program, or you can use a program designed strictly for creating separations.

The Final Save

Now that all of your post-processing is done, it's time to save the image. You may want to get into the habit of saving the image after each step—spotting for dust, the final color adjustment, and unsharp masking. When you save the image after post-processing, make sure the file format you choose is compatible with the next stage of production. See Chapter 7, "Saving and Storing Images," for details.

Conclusion

Scanning is a complex process that involves many steps. Every step, from properly placing the scanner, to saving the image after post-processing, is important to the final look of the image. Once the process becomes routine, you'll be able to efficiently produce high-quality scans every time.

C H A P T E R

6

Understanding and Using Color

Understanding Bit-Depth
Color Separations

Color is one of the more challenging aspects of computer graphics and print reproduction. But there is an underlying logic to it all. Understanding how scanners acquire color and how color is reproduced will enable you to more accurately control the color and detail of your scans. Because this topic is not simple to grasp, this brief chapter cannot make you a color expert, but it will give you the foundation you need to experiment in a controlled, logical way.

Understanding Bit-Depth

You may be wondering what bit-depth has to do with scanning and adjusting images. The pixels that compose an image are controlled by one or more switches, or bits. The number of bits assigned to control each pixel is known as the image's *bit-depth*. The more bits assigned each pixel, the greater the possible number of shades of color that are available to a pixel.

One-Bit Images

One of the first things you learned about computers was probably that they break down all information into 1s and 0s. A bit determines whether a piece of information is represented by a 1 or a 0. If the bit is switched off, that information is represented by a 0; if it is switched on, the information is represented as a 1.

In a 1-bit image, only one switch is assigned to each pixel. If the switch is on in an image, the pixel is white; if it's off, the pixel is black, as depicted in the illustration that follows. Therefore, in a 1-bit image, a pixel can only be either black or white. No shades of gray are possible. To see what a photograph looks like when represented as a 1-bit image, see the plate "Color Depth" in the color section of this book.

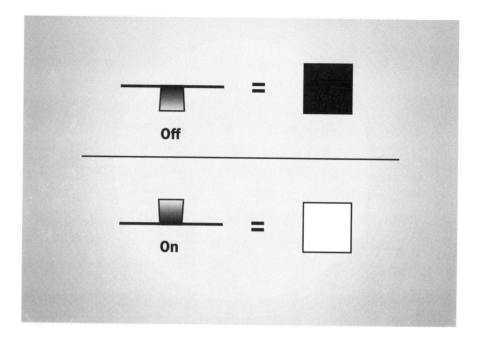

The next section discusses how additional bits produce grayscale images.

Grayscale Images

In a grayscale image, each pixel can be a shade of gray, it can be black, or it can be white. These alternatives are possible because more than one bit controls each pixel. As more bits are used to control each pixel, the number of possible combinations grows exponentially. For example, if two switches control each pixel, there are four possible combinations, as shown in the following illustration.

If the number of bits that control each pixel is increased to eight, there are 256 possible combinations, and therefore 256 different levels of gray. However, most people can't see the subtle differences between neighboring tones when 256 levels of gray are present. Therefore, an image that can represent 256 different shades appears

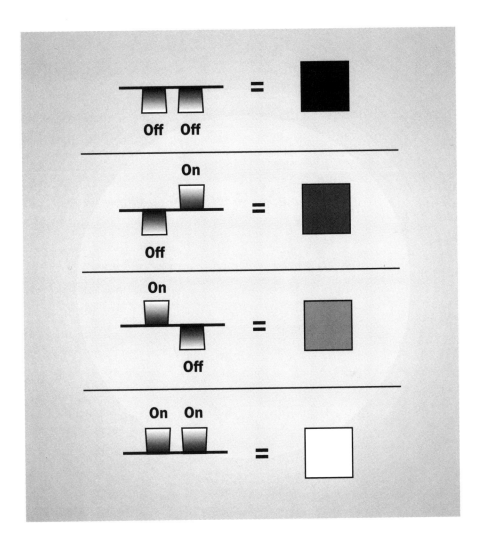

to have a continuous tone. Because it takes eight bits to create 256 levels of gray, a photo-realistic grayscale image has a pixel depth of eight bits.

Color Images

It's natural to assume that color originals are scanned in color because an original is placed in the scanner. Then, after it's scanned, the image appears in color on the monitor. In fact, images are not scanned in color at all.

A scanner captures an image as three separate grayscale channels. It does this by reading the intensity of light that passes through the original in the case of a transparency scanner, or reflects off the original in the case of a flatbed scanner. The light then passes through red, green, and blue (RGB) filters. These grayscale channels are then assigned the color that corresponds to the filter through which they were acquired. When the grayscale channels are given their appropriate color, the shades of gray are replaced with shades of color. The darker the areas of the grayscale image, the less color it gets. In order to represent the widest possible range of tones for the RGB component colors, each of the red, green, and blue channels should have a bit-depth of eight bits. The colored channels are then merged together and the color image is created.

When the three channels are merged, the eight bits that control each pixel in each channel are combined. As a result, each pixel of the color image is controlled by 24 bits (3×8). Because the total possible on and off combinations of 24 switches is 16.7 million, in a 24-bit image, each pixel can be any one of 16.7 million colors. This far exceeds the number of discrete colors that anybody can discern, making a 24-bit image appear as continuous tone color.

Why Do We Need More Than Eight Bits?

If eight bits of information for each pixel creates 256 shades of gray or color, and most people can't see more than 256 shades, why can some scanners recognize 10, 12, or even 16 bits per pixel? There are three situations in which the ability to acquire more than eight bits per pixel is useful: to get better shadow detail, to allow more control over the tone curve, and to remove color casts. These topics are discussed below.

Getting Better Shadow Detail

As discussed in Chapter 3, scanners introduce noise, or speckles, into the shadow areas of an image when converting light into digital information. Scanners will normally filter out the noise, but some shadow detail gets filtered out as well. If a scanner only acquires eight bits of information per pixel, and then filters out some tonal steps to remove the noise, you're left with substantially less than the 256 steps required to reproduce the image's full tonal range. However, if a scanner can acquire 10 or more bits of information per pixel, there will still be eight bits of usable data for each pixel even after the noise is filtered out. This will leave enough tonal range to reproduce all the necessary detail.

Control Over the Tone Curve

Scanners which can capture 10 or more bits of information from an original give you the freedom to select the tonal area you wish to preserve. These scanners allow you to stretch out the tonal curve in the highlight, midtone, or shadows in order to maximize the number of tonal steps in that area. This gives you much more control over the detail in the original, so detail can be preserved all the way through to the printed piece.

Removing Color Casts

The ability to acquire more than eight bits per pixel is also useful if you need to scan an original with a severe color cast. To make a heavily casted original look natural, some color information must be removed. If you scan a casted original on a scanner capable of acquiring only eight bits per pixel, you can have two options. You can either adjust the scanner so it won't acquire the full amount of the cast color, or scan the image without compensating for the cast, removing it afterwards with image editing software. In either case, you start with 256 possible shades for the cast color. Then you either remove, or choose not to acquire, some color to correct for the cast. As a result, you end up with less than the full possible tonal range for the cast color, and therefore hamper your ability to hold detail in that color. To see a comparison of a casted original before and after it is corrected for the cast, see the plate "Color Cast" in the color section of this book.

If a heavily casted original is scanned on a scanner capable of recognizing 10 or 12 bits per pixel, you can compensate for the cast with software controls before the scan is made. A scanner that can acquire the entire tonal range as 10 or more bits of information for each color can read thousands of shades of each color. Adjusting this type of scanner to pick up less of the cast color still enables you to acquire eight bits of color information, or 256 shades, for the cast color. Therefore, you will be able to reproduce the entire tonal range.

Scanning Poorly Exposed Originals

You'll also get much better results from an improperly exposed original if you use a scanner that has a bit-depth of 10 bits per pixel or greater. A poorly exposed original may have all of the important detail defined by very subtle steps within a narrow tonal range. A scanner with a bit-depth of eight bits per pixel allows for only 256 tonal steps for the entire image. This means the scanner may not pick up all of the important detail if it's concentrated in a narrow range. A scanner that can acquire more than eight bits per pixel can recognize thousands of shades, so it is able to acquire the 256 shades required for smooth tonal reproduction even from a relatively narrow tonal range.

Remember that even the best scanners can only do so much with a poor original. If the original is so poorly exposed that you can barely see the detail, no scanner will be able capture it for you.

Color Separations

All color scanners acquire color images by breaking them down, or separating them into their red, green, and blue (RGB) components. Because monitors use the RGB color model to display images, this model is fine if you only plan to display the image on a video monitor. However, if the image will be printed, it needs to be converted from the RGB color model to the CMYK (cyan, magenta, yellow, and black) color model.

Why RGB and CMYK?

There are many differences between the way monitors and paper present images. One of the most important differences pertains to the color of the background. For example, if a monitor is not displaying anything, the screen is black. Because black is the *absence* of light, we must add light in order to create images on a monitor. We use the RGB color model to create colors on a video monitor because adding red, green, and blue light together, in full and equal intensities, creates white. This is why red, green, and blue are known as the *additive colors*. Mixing together various combinations of these colors will give us virtually any color in the visible spectrum. To see how red, green, and blue light combine to make white, see the plate "Color Models/RGB" in the color section of this book.

When printing, however, we start out with white paper instead of the black background of a monitor. We see paper as white because it reflects back all the light that hits it. Because white is the *presence* of all colors of light, we create different colors by subtracting color from white light. Therefore, to create color on white paper, we need to filter out, or subtract, some of the light reflected back from the paper. This is done by applying cyan, magenta, yellow, and black (CMYK) ink. This explains why cyan, magenta, yellow, and black are known as the *subtractive colors.*

Viewing Color Separations and Channels

Most separation software lets you view the CMYK image on your monitor. It is important to remember that you're viewing only a simulation of what the CMYK image looks like because video monitors can only display in RGB. In order to get a more accurate view of a separation on screen, you should look at each of the channels on screen individually.

Also keep in mind that when viewing RGB channels on your monitor, the more intense the color in an area, the lighter that area will be when displayed on a monitor or printed. This is because in the RGB color model, colors are added together to make white. When viewing CMYK channels, the more color (or grays), the darker the areas will be. This is because in the CMYK color model, more color means more ink, which reduces the amount of light reflecting off the paper. To see what separate channels look like in both the RGB and CMYK color space, or color model, see the plates "Color Models/RGB" and "Color Models/CMYK" in the color section of this book.

When to Create Separations

All methods of printing on paper use the cyan, magenta, yellow, and black pigment of some sort. The pigment is either ink, dye, or toner, depending on the type of device. Therefore, any image to be printed needs to be separated because it was acquired as RGB. If your scan will be printed on an offset press, you need to either do your own separation, or perhaps the service bureau that outputs your film will do it for you, for a fee.

If you're going to print your image to a digital color printer, the specific printer may require that the file be sent to it as CMYK, in which case you'll need to make the separations yourself. However, some digital printers will separate RGB files that are sent to them. This can be beneficial because separations are made based upon the characteristics of the output device. As a result, a printer can very accurately separate an RGB image to look its best on that device. Sometimes, however, you will have more control over the color of an image if you do the separation yourself.

Creating Separations

Most image editing software, and even some page layout programs, will separate scanned images. The level of control and the quality of the results vary widely. There are also applications that are used only for creating separations.

Bad separations stick out like a sore thumb. No matter what type of software you use to create separations, you need at least a basic understanding of the separation process in order to produce the best results. What follows is a brief explanation of what happens when an image is separated and how to control this process. You can expect to find these controls in almost any software that can create separations. However, software designed specifically for color separation exists and gives you an even greater degree of control.

The Challenge of Black

RGB and CMY are perfect complements of each other. Removing red makes an image more cyan, removing green makes an image more magenta, and removing blue makes an image more yellow. As mentioned earlier, adding together red, green, and blue light in full and equal intensities will produce white. It follows that adding together equal amounts of cyan, magenta, and yellow should produce black. While this may be true in theory, mixing these three colors together in full intensity will yield a muddy brown color, due to the inherent impurity of printing inks. Therefore, the only way to create true black when printing is to add black ink. Deciding how to take information from the cyan, magenta, and yellow plates to create the black plate, known as black generation, is the science and craft of color separations.

Gray Component Replacement　There are two primary ways of dealing with black generation when separating an image: GCR (gray component replacement) and UCR (under color removal). In a four-color image, some areas will be made up of a combination of cyan, magenta, and yellow ink. Mixing equal amounts of cyan, magenta, and yellow ink creates a neutral gray. Even if the colors are present in unequal amounts, as long as cyan, magenta, and yellow are all present, there will be neutral component to the color. This neutral component is based on the lowest common percentage of all the colors. GCR replaces some of this common amount of cyan, magenta, and yellow with black. For instance, a deep blue may be composed of 100% cyan, 20% magenta, and 60% yellow. If GCR were used to generate black, depending on the level of GCR used, perhaps 10% of each color would be replaced with black. As a result, the color would now be made up of 90% cyan, 10% magenta, 50% yellow and 10% black. This would mean that not all of the neutral components would be removed. The Black Generation setting allows you to determine where in the tonal range, and to what degree, black will replace colors.

　A GCR setting in Photoshop looks like this:

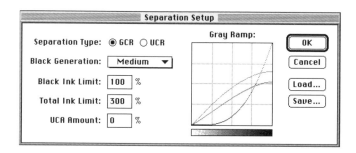

Under Color Removal　UCR is the other common black-generation technique. When cyan, magenta, and yellow are present in equal amounts, UCR replaces these colors with black. As mentioned in the discussion of GCR, equal amounts of cyan, magenta, and yellow ink cancel each other out, creating a neutral gray tone. UCR reproduces the neutral areas with only black ink. A UCR level can be set so it doesn't kick in until the ink reaches a certain density. This has the effect of only replacing color with black in the midtones and shadows, which is desirable because black ink can make the highlights look dirty. UCR is a good method to use when printing on uncoated paper stock because it drastically reduces the amount of ink applied to, and absorbed into, the paper.

A UCR setting in Adobe Photoshop looks like this:

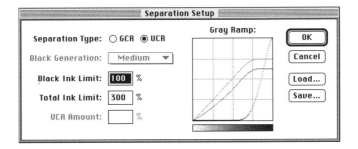

Other Separation Settings

As important as black generation is, there are other things to consider when setting up a separation. These settings not only affect how the image looks in print, but may also cause printing problems if not set properly.

Total Ink Limit Another factor to consider when making separations is the total ink limit. This setting determines the maximum amount of ink allowed in the densest part of an image. If the total ink limit is set for 400%, the densest area of an image can be made up of 100% of each cyan, magenta, yellow, and black. Very few presses can handle this amount of ink, however. Additionally, the paper will become saturated and the ink will smear because it won't have time to dry. If printing is done on a web press, the giant rolls of paper may tear if too much ink is applied. Typical settings are 320% for a sheetfed press and 280% for a web press. The actual total ink limit depends on a variety of factors, including the type of paper, inks, and press used for the print job. Be sure to consult your printer before using these settings.

Under Color Addition   When a large solid black area is created with black ink only, it looks flat. The UCA (under color addition) option allows you to add a designated percentage of cyan, magenta, and yellow under solid areas of black to give them a richer, more saturated look. UCA is not available when using UCR because UCR is designed to remove any ink that is not black from neutral areas.

Printing Inks Setup   Every type of color output can reproduce its own unique range of color, which is known as the device's color gamut. The RGB color model that your scanner and monitor use has a much wider gamut than what's available with most types of printing. Therefore, part of the separation process is to limit the color gamut of the image to match that of the final output device. This means before you create a separation, you need to let the software know what type of device you'll be printing to. This is very important because if the incorrect gamut is set, the separation will either contain color that the output device can't reproduce, or the output will be printed with a narrower range of colors than the printer is capable of reproducing.

You choose the color gamut for a separation by selecting the printer type in your separation software. If you'll be using a digital printer and don't see it listed in the area of your software where you select the type of printer to be used, you may be able to get the setting from your printer's manufacturer. Some software allows you to create your own gamut settings, but this takes a fair amount of expertise to do successfully.

The Printing Inks Setup dialogue box in Photoshop looks like this:

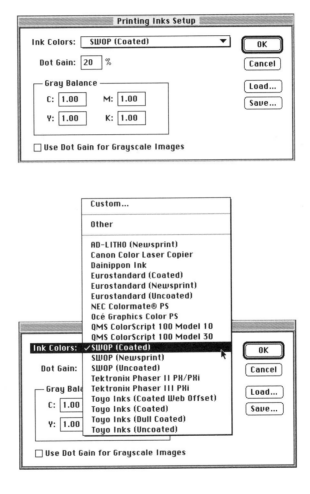

Specifying the printing device in the Printing Inks Setup dialog box tells the software what the range of reproducible colors will be when converting from RGB to CMYK.

Conclusion

Understanding color, on the computer and in printing, is the key to gaining control over the way your images are reproduced. Knowing how scanners turn light into computer data will enable you to acquire all of the necessary detail in an original. Understanding how the RGB image created by the scanner is converted to the CMYK color model used for printing enables you to make informed color decisions. You're now on your way to getting important detail and consistent, pleasing color from your original.

C H A P T E R

Saving and Storing Images

File Formats
Compression
Storing Images

The last stage of scanning is saving the image. This involves several choices, including determining the file type, or format, and deciding whether or not compression should be used. The correct format is determined by the subsequent steps in the production process. If you'll be printing the scan by itself, the file format must be compatible with the target printer. If the scan will be used in a page layout document, slide presentation, or multimedia project, you need to choose the file type that works best with the particular program that will integrate the scan. The type and level of compression, if compression is to be used, will depend upon your needs and capabilities for storing, transporting, and printing images.

File Formats

Not very long ago, one of the most confusing aspects of computer graphics was sorting through the alphabet soup of file formats. Most applications could only work with images in a specific format and converting images between formats was cumbersome. This is less of a problem today because graphics programs can accept images in a variety of formats and the major image editing programs can easily convert images between file formats.

There are many different types of file formats (see Figure 7.1). While some are native to specific applications, most can be used with many different programs, even across different platforms. This section covers the file formats most commonly used for scanned images. All of the major image editing programs can save images in these formats.

Figure 7.1
The Save As dialog box in Photoshop gives you an idea of the variety of file formats that are available.

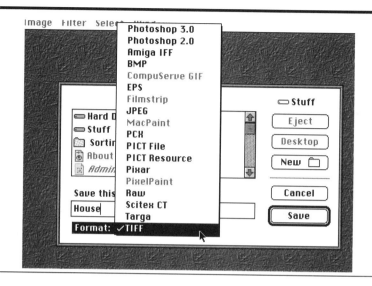

TIFF: The Tagged Image File Format

A TIFF image can be either bitmap (line art), grayscale, RGB color, or CMYK color. The most recent revision to the TIFF standard, TIFF 6.0, also allows you to save spot color information. Most page layout programs will allow you to import images as TIFF files. These programs can print placed TIFF files as composite color, meaning that all of the color information in the file can be printed together. Some programs will also separate a CMYK TIFF file, so when creating separations, the cyan, magenta, yellow, and black channels print on separate film plates. Many presentation and multimedia programs don't accept TIFF files, so be sure to check the application's manual before you save files for importing into these programs.

TIFF is a relatively efficient format, so an image saved as a TIFF file doesn't take up as much disk space as some other formats. TIFF files can be made even smaller by using LZW (Lemple-Ziv-Welch) compression, a type of compression only available with TIFF files. (For more on LZW, see "Compression," later in this chapter.)

TIFF files work equally well on both the Macintosh and Windows platforms (see Figure 7.2). There are some differences between the way the TIFF standard is implemented on each of these platforms, but an image editing program on either platform can open virtually any TIFF file. If the image will be placed into a page layout program, you may first need to open it in an image editing program and save it as a TIFF file that is native to that platform.

Figure 7.2
Photoshop allows you to save TIFF files that are compatible with either the Macintosh or Windows platforms, and gives you the option of adding LZW compression.

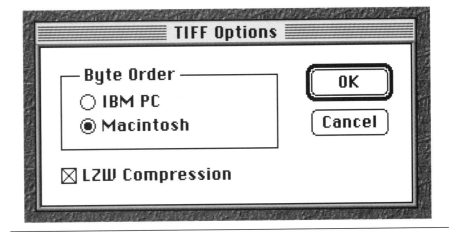

The original specifications for the TIFF format were somewhat loosely defined. As a result, each application implemented it slightly differently, and compatibility problems arose. In the past few years, all applications have been using the same version of TIFF, so compatibility shouldn't be a problem unless you're using software that is several years old.

EPS: Encapsulated PostScript

Like TIFF, EPS also exists on both the Macintosh and Windows platforms. And like TIFF files, EPS files can be used for either bitmap, grayscale, RGB, or CMYK images as well as duotones, tritones, and quadtones.

PostScript is the standard language printers use for creating images. Unfortunately, your computer can't directly display PostScript.

In order to view EPS files on your monitor, the images need to be saved with a preview (see Figure 7.3).

Figure 7.3
When saving an EPS file, you have the option of creating a preview for the Macintosh or Windows, and either color or black and white. You also have the option of not creating any preview at all.

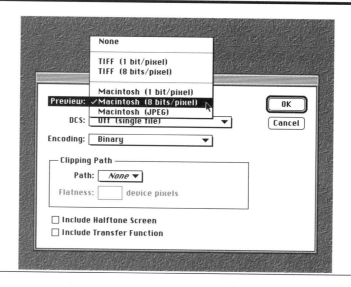

There are several different types of previews to choose from. Selecting None will save the image without any preview. If you choose this option, when the picture is placed in a page layout program, you won't see the images, but will instead see a gray box that says PostScript File and the name of the file. Saving the image without a preview will create a smaller file than if you save with a preview (see Figure 7.4). However, it's almost impossible to accurately place, crop, and size an image in a page layout document without a preview. Exactly how much smaller a file without preview will be depends on the physical size of the image. Previews of EPS files are created to the true physical dimensions of the image, at a resolution of 72 ppi. The larger the physical size of the image, the larger the file size will be if a preview is added. The amount that a preview will add to the file size is usually negligible, unless the image is for poster-sized output.

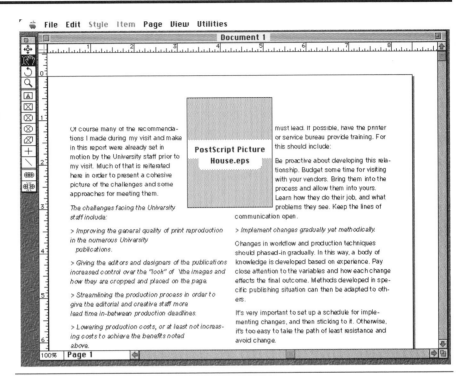

Figure 7.4
If an image is saved as an EPS file without a preview, and you import the picture into a page layout program, this is what you'll see.

In most cases you'll want some type of preview so you can actually see the image when it's imported into another program. If you'll be using the image on the Windows platform, you'll want to choose one of the TIFF preview options. If the image will be used on the Macintosh, then obviously one of the Macintosh preview options should be chosen. When one of the Macintosh options is chosen a PICT preview is created. The Macintosh options can only be used on the Macintosh because PICT files can only be used on the Macintosh platform.

A TIFF or Macintosh preview can be either one bit or eight bits. The 1-bit preview is bitmapped, which uses only black and white pixels. The 8-bit previews are 8-bit color. While 1-bit previews create smaller files than 8-bit previews, you'll want to use the color previews unless the picture has a very large physical dimension or disk space is very limited. If you're working on the Macintosh platform, you can also use a JPEG preview (see "JPEG Compression," later in this chapter). This is a color preview that is compressed to take up less disk space. JPEG is a relatively new type of preview option, and it may not be compatible with all other applications.

When choosing between these preview options it's important to remember that they affect only the preview of the image, not the actual image data. The image will print the same no matter what type of preview you choose.

DCS: Desktop Color Separation

If you're saving a CMYK image as an EPS file, you can choose whether or not to save them as a DCS (desktop color separation) file (see Figure 7.5). If DCS is set to Off, a single file is created that holds the preview as well as the image data for the cyan, magenta, yellow, and black plates. Choosing one of the On options creates five individual files by separating the preview and each of the four color plates into discrete files. Single-file EPS files are a bit easier to work with because

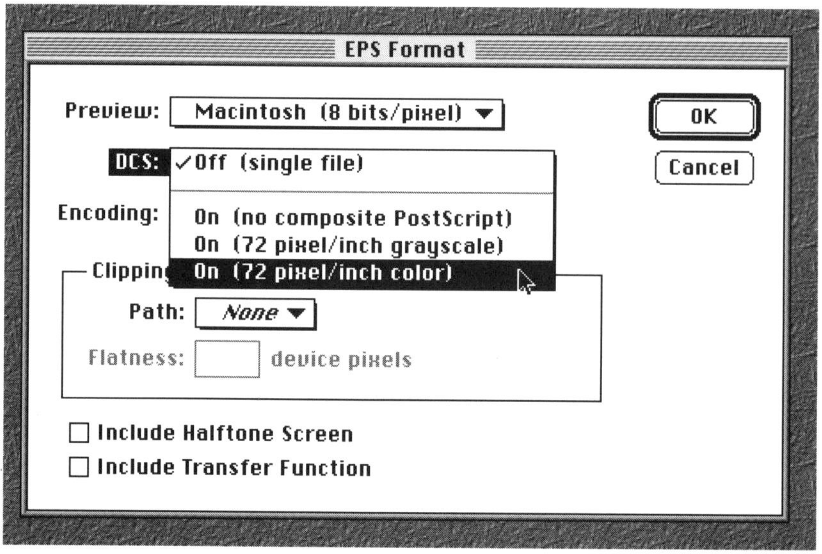

Figure 7.5
Saving an image as a DCS file allows you to create separate files for each plate and the preview.

EPS Format

Preview: Macintosh (8 bits/pixel) ▼

DCS: ✓Off (single file)

Encoding: On (no composite PostScript)
            On (72 pixel/inch grayscale)
Clipping   On (72 pixel/inch color)

Path: None ▼

Flatness: [   ] device pixels

☐ Include Halftone Screen
☐ Include Transfer Function

OK

Cancel

there are less files to keep track of, but they also print slower than DCS files. In general, the larger the image file, the greater it will benefit from being saved as a DCS file.

Another nice thing about DCS files is that by freeing the preview from the image data, designers can avoid the pain of working with large images if they are simply placing, cropping, and sizing the image in a page layout program. The preview is usually much smaller than the image files, so time is saved by not having to move large images around via disk or over the network. As long as the preview and image files are reunited in the same folder at print time, or are relinked through the page layout program, the image will print using the high-resolution image data. There are some printers that can only process composite color information. These printers won't use the high-resolution data, but will instead print the low-resolution preview.

Encoding

There are several ways to encode EPS files (see Figure 7.6). The two primary options are ASCII and binary. In most cases you'll want to

Figure 7.6
Encoding EPS files

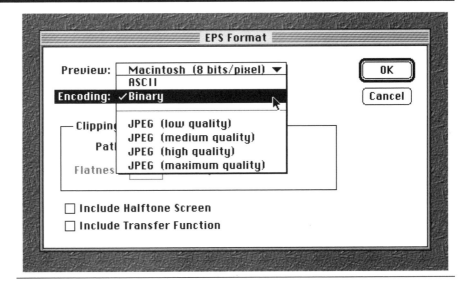

choose binary. Once upon a time, some page layout programs and printers could only accept EPS data if it was ASCII encoded. The problem with ASCII is that it creates files about twice the size of binary encoded files. This makes ASCII encoded images slower to print and more expensive to store. Unless you're using an older page layout program or printer, you'll want to avoid ASCII.

It is also possible to encode EPS files with JPEG compression. JPEG creates smaller files but at the cost of some loss of image quality. If the biggest bottleneck to printing speed is the time it takes to send the image over the network to your printer, you may benefit from the smaller files created by JPEG encoding. Whether the loss of quality will be visible or not depends on the JPEG setting chosen and the quality requirements of your output. Only PostScript Level 2 printers can print JPEG-encoded EPS files. Older printers don't have the capability to decompress the JPEG files and will likely crash. (JPEG compression is covered in greater detail later in this chapter.)

Choosing between TIFF and EPS

Most page layout programs can accept scanned images as either TIFF or EPS files. While TIFF files are more compact and more reliable, there are some situations in which it is advantageous to use EPS. When deciding which format to use, there are several things to consider, such as screening information and clipping paths. Read on for details.

A Few More Options

It is possible to include a clipping path when saving an EPS file. Clipping paths are created so an irregular (nonrectangular) portion of the image can be brought into another application. (Clipping paths are covered in more detail later in this chapter.) If you decide to use a clipping path, you can choose to give it a level of flatness. A setting of 0 flatness forces the path to print in the exact shape in which it was created. Raising the level of flatness allows the path to

cut corners, so to speak, thereby making the path a bit less accurate. The higher level of flatness also makes the file much easier for the printer to calculate, and therefore print. For most paths, a flatness of 3 will not create any visible difference and may drastically reduce printing times.

Include Halftone Screen and Include Transfer Function are other options that are available only when saving a file in EPS format. If halftone screens were set in the image editing programs, Include Transfer Function must be selected in order for the screening information to be included with the image. You should only choose Include Transfer Function if the printer was calibrated through the image editing program and not through separate printer imagesetter-calibration software. (For more on halftone screens, see Chapter 4, "Resolution," and Appendix A, "Halftones." For more on calibration, see Chapter 8, "Calibration.")

Screening Information

In the early days of color desktop publishing, proper screening was one of the biggest challenges that was faced when reproducing color photos. Today, most printers have screening information built in to their RIPs (Raster Image Processors). The RIP is the part of the imagesetter that translates the PostScript language (sent by the computer) into the dots that are printed. As a result, in most cases you won't want to include screening information along with your image. If, however, you'll be printing to an imagesetter that does not add screening information, or you want to use custom screening patterns to create special effects, you'll want to use EPS format.

While TIFF files can't contain any screening information, most page layout programs will allow you to add screening information to a TIFF file after it's been imported. Unfortunately, most page layout programs do not give you as many options for setting screening as does image editing software. As a result, you may not have as much control over the screening of TIFF files as you do over EPS files, which get their screening from image editing programs.

Clipping Paths

Another reason you may want to use EPS files instead of TIFF files is that Adobe Photoshop, the most popular image editing program, allows you to use a clipping path if an image is saved as an EPS file. Clipping paths are used when an image is exported as an irregular shape. Normally, when you save an image in an image editing program, the shape of the image must either be square or rectangular. Even if the background has been deleted in order to silhouette a subject, when the image is placed in a page layout program, the subject will be surrounded by a white rectangle. By enclosing the subject in a clipping path, only the subject will be exported and the irregular shape is maintained. This is very useful if you need to silhouette a product shot in a catalog or ad.

Other File Formats

TIFF and EPS are the most popular formats for scanned images because they are accepted by most printers and most desktop publishing applications. However, there are many other file formats. The ones you're most likely to encounter are covered in the next few sections.

PICT

The PICT format exists only on the Macintosh platform. PICT files can be either bitmap, grayscale, or RGB, but not CMYK. PICT files have their own built-in compression scheme, making them small compared to the same image saved as a TIFF or EPS file. PICT is not considered a reliable format because it is not well defined, and may print unpredictably. Some applications, mostly presentation and multimedia programs on the Macintosh, will only accept imported images as PICT files.

Photoshop Format

As the name suggests, Photoshop format is the native file format of Adobe Photoshop. Photoshop format files can be either bitmap, grayscale, RGB, or CMYK. Files in the Photoshop format have the following advantages over other formats:

- They save and open more quickly than other formats.

- They are easily transported across platforms, even to UNIX workstations running Photoshop.

- Unlike almost all other file formats, Photoshop files can also hold additional channels of information, which can be used to store masks and selections.

Most desktop applications will not accept Photoshop format files as imported images. But Photoshop format is the way to go, if any of the following criteria apply:

- You use Adobe Photoshop.

- You open and save an image several times before importing it into another program.

- You need to move an image to another platform.

- You want to do some complex image enhancement.

PCX

PCX was originally designed for paint programs run under DOS, and later Windows. Previous versions were only capable of 8-bit color or less, but the most recent version is capable of 24-bit color. Images can be in RGB color, but not CMYK separations. There are very few instances when you would actually want to save a scan as a PCX file instead of as a TIFF or EPS. The rare exception may be the need to integrate a scan into an older version of a DOS or Windows presentation program.

Compression

Scanned images are some of the largest files most of us will encounter. Word processing, spreadsheet, and even page layout documents rarely grow to larger than a few hundred kilobytes. Conversely, even small scanned images can be several megabytes, with larger images easily exceeding 10MB, and some weighing in at 50MB or more. Storing and transporting scanned images can be extremely expensive. As a result, compression is particularly attractive to those who deal with scanned images.

Compression reduces the amount of disk space required to store an image or another type of data. Compression can either be lossless, in which no original data is lost, or lossy, in which some of the original information is lost or replaced. While lossless compression may be used for either images or other data—such as word processing documents or spreadsheets—lossy compression is used exclusively for images.

Lossless Compression

As the name implies, lossless compression reduces the size of a file without losing any information. The amount of compression is usually pretty moderate, but no image quality is sacrificed in the process. Disk compression software reduces the amount of disk space required to store an image by making more efficient more use of the storage medium. Common programs for the Macintosh are StuffIt, CompactPro, and DiskDoubler; PKZIP and Stacker are widely used for Windows applications.

Disk Compression

While this method can yield significant savings for some types of data, image files usually yield 2:1 file compression at best. Sometimes savings may be less than 10 percent.

There are two major disadvantages of saving an image using loss-less compression, aside from the relatively small amount of file size savings. The first disadvantage is that opening and closing compressed files takes longer than opening and closing uncompressed files. Saving a few hundred kilobytes of disk space may not be worth having to wait several minutes each time you want to open an image.

Another disadvantage of lossless compression is the files tend to be harder to recover from corrupted hard drives. Hard drive failure happens more often than most of us like to think. There are several utilities on the market that are designed to rescue files from crashed drives. Their rate of success with compressed files of any type is substantially lower than with uncompressed files.

LZW Compression

When saving an image in TIFF format, most image editing programs give you the option of saving the file with LZW (Lemple-Ziv-Welch) compression. Because this type of compression is lossless, the quality of the image is preserved. If you'll simply be storing your images for your own use, this is a good reason to take advantage of LZW compression, although it may take a bit longer to open and save your images. If you plan to import your images into another program, check to see if that program can accept LZW compressed files. Most of the newer programs can. Some printers, but not all, can print LZW compressed images directly, without requiring that you first decompress them on the computer. Again, it is mostly the newer printers that can do this.

Lossy Compression

With lossy compression, very high levels of compression are possible, but at the cost of losing some image quality. Whether the loss of quality is visible or not depends on the content of the image, and its resolution.

JPEG Compression

The most common method used for compressing images is JPEG (Joint Photographic Experts Group). JPEG can compress files to between $\frac{1}{2}$ to $\frac{1}{40}$ of their original size.

While primarily used as a lossy method of compression, the JPEG standard includes a lossless compression method. When used for lossless compression, JPEG will reduce the file size by about $\frac{1}{2}$.

When using JPEG, there is always a trade-off between the level of compression and image quality. The more a file is compressed, the worse the image will look when it's decompressed. (See Figure 7.7.)

Figure 7.7
When saving an image with JPEG compression, the higher the image quality, the lower the level of compression.

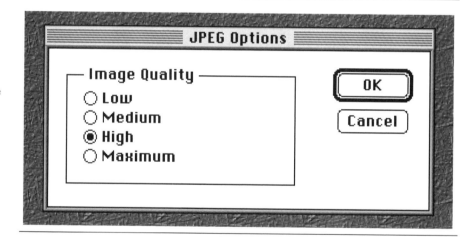

JPEG works by throwing away redundant color information. This means if the color value of one pixel is similar to that of other pixels in the image, only one color value is "remembered" for all of these pixels. When the image is decompressed, this one value is then given to all of the pixels that had similar values. The higher the level of compression used, the greater the range of values that are considered similar. To see how different levels of compression affect printed output, see the plate called "JPEG Comparison" in the color section of this book.

The content of the image also affects how much a file will get compressed. An image made up of solid colors won't have many unique pixels, so JPEG's method of using the value of a few pixels to represent all of the image data can be very efficient and not result in any visible image quality loss. On the other hand, a complex and detailed image will have many more unique pixel values. In this case, a high rate of compression is more likely to result in a loss of image quality.

JPEG degrades a high-resolution image much less than a low-resolution image. This is because JPEG evaluates and replaces pixels in an 8×8 pixel matrix. The higher an image's resolution, the more likely it will have identical or very similar pixels neighboring one another.

Should You Use JPEG?   There are many circumstances when JPEG compression can be very useful. If you need to transmit an image via modem, compressing an image at a high-quality setting may not make any visible difference in the picture, but will cut transmission times significantly. This can be especially important when trying to deliver an image to meet a tight deadline.

JPEG can also greatly reduce the cost of storing images by allowing you to fit more images on a hard disk or other storage medium. Be careful not to compress the images too much or they won't look good when they're ultimately used. Like most forms of compression, one disadvantage is it takes longer to open and close files saved as JPEG than it does for uncompressed files.

JPEG compression may also allow you to print faster. Printers that use PostScript Level 2 can automatically decompress JPEG files. Because JPEG files are smaller, they are sent to the printer faster, which can greatly reduce the amount of time needed to print large images.

Recompressing JPEG Files   Once you choose a level of compression for an image, it should always be recompressed at the same level. Using the same compression level over and over again will not increase the amount of data lost above what was lost the first time the data was compressed. But if an image is compressed and recompressed repeatedly at different levels each time, the image will quickly break down and become unusable.

Storing Images

Scanning creates some of the largest files you are likely to encounter on a personal computer. Even a relatively small image can take up several megabytes of disk space, and larger images may weigh in at 20MB or more. With files of this size, even a few images can quickly fill up your hard drive. While you may not want or need to save every image you scan, you will want to save some of them. At the very least, you'll need a place to store and back up working files.

When considering what type of storage medium to use, the primary considerations are cost and convenience. Factors that affect convenience include, the performance of the medium, its compatibility with other sites you want to exchange data with, and the ease with which you can access your images.

Unless you plan on saving only a few images and don't plan on exchanging files with anyone else, you need to use two types of storage: a big, fast hard drive, to give you room to work on large images and fast access to images you frequently need; and some type of removable medium for backing up your files, transporting them, and storing them more economically.

The following sections summarize your storage options.

Hard Drives

When personal computers were primarily used for word processing, number crunching, and creating simple bitmapped graphics, a 20MB hard drive was sufficient and an 80MB drive was ample.

Today, if you scan and edit color images, a 270MB drive is barely adequate and a drive with a capacity of 1 gigabyte (1000 MB) or more is preferable.

When working with scanned images, the performance of the drive is extremely important. Because scanned images tend to be so large, a slower drive takes noticeably longer when opening, copying, or saving an image. The speed of the drive is even more crucial when using virtual memory, which moves the image between the processor and the drive as you're working with it. For a more detailed explanation of virtual memory, see the sidebar in this chapter, "Virtual Memory."

While the price of hard drives has dropped significantly in recent years, it still costs more to store images on a hard drive than on removable media. As a rule, the larger the drive's capacity, the cheaper it is to store each megabyte of data.

A hard disk will always be your primary type of storage medium, so you should get the biggest, fastest drive you can afford.

Removable Storage

Eventually, your hard drive will fill up. At that point you can buy yet a bigger hard drive, but that's a very expensive way to increase storage capacity. A better solution is to use a removable storage system. Removable storage systems have two parts—a drive unit that houses the read/write heads, power supply, and connection terminals; and a cartridge, disk, or tape, where the data is actually stored. You only need one drive unit to access a limitless amount of removable media. Removable storage is more cost efficient than hard drives because whenever you need to increase your storage capacity, you only need to buy the media, and not all the supporting hardware that you only pay for once as part of the drive unit.

Figuring out the cost of using a particular type of removable medium can be a little tricky because the price of the drive unit

Virtual Memory

The processor in your computer can only access information that is in RAM (random access memory). RAM is fast, but it's also about 50 times more expensive per megabyte than hard disk space. Wouldn't it be nice if you could use inexpensive disk space instead of expensive RAM? Well, you can, through a technique known as virtual memory. Virtual memory tricks your computer into thinking that part of you hard disk is actually RAM. This allows you to keep more applications open and work on files that are larger than you could fit into RAM.

Virtual memory is especially important when working on large images because they will often be too large for the systems that use true RAM. Without virtual memory, a system cabable of editing large images would be prohibitively expensive because RAM is so pricey. But the cost savings come at the price of performance. Your hard disk is not nearly as fast as real RAM. Most image editing software requires that you have three times as much RAM as the size of the image you're working on. This means that if you're working on an image that is larger than one-third the amount of RAM you have available, the image data will constantly need to be copied back and forth between the RAM and the hard drive. Getting a drive with the fastest performance you can afford is the best way to take advantage of the cost savings of virtual memory, while minimizing the performance penalty.

must be factored in with the cost of the medium. For more on calculating the cost of storing data on removable media, see the sidebar in this chapter, "The Cost of Removable Storage."

The following sections take a closer look at some of the most popular removable storage systems.

SyQuest

SyQuests are similar to hard drives, except the platter that holds the data is in a removable plastic case, while the read/write heads are in the drive unit.

SyQuest cartridges come in two sizes, the original $5^1/4$ inch and the newer $3^1/2$ inch. The $5^1/4$-inch system comes in capacities of 44, 88, and 200MB. The $3^1/2$-inch system is available in 105 and 270MB capacities. Each capacity uses a different drive and different cartridges. Within each size, the cartridges are upwardly compatible so you can use a lower capacity cartridge in a higher capacity drive but

The Cost of Removable Storage

How much does it cost to store an image? The easiest way to figure this out is to calculate the cost per megabyte of your storage system. With hard drives, this is fairly easy to do. Just take the cost of the drive and divide it by the drive's capacity. For example, a 500MB drive that costs $300 has a cost per megabyte of $.60.

This gets a bit more complicated when trying to calculate the cost for removable media, such as a SyQuest, Bernoulli, or M/O (magneto optical). That's because the cost of the drive unit must be factored in with the cost of the medium. As a result, the more data you store, the lower the cost per megabyte.

For example, an 88MB SyQuest drive costs approximately $300 and each cartridge costs around $75. If you're storing 880MB of data, you can find the cost per megabyte of storage by adding together the cost of the drive ($300) and the cost of the 10 cartridges ($750) and dividing the total by 880. The cost per megabyte is 1.19 ($1,050 ÷ 880).

However, if you'll be storing 8.8 gigabytes (8,800 megabytes) of data, then you'll need 100 cartridges. To find out the cost per megabyte, the price of the drive ($300), and the price of 100 cartridges ($7,500) are added together, and divided by 8,800. The cost per megabyte now, drops to 88 cents ($300 + 7500 = 7800 ÷ 8,800 = .88).

The example above assumes you only need one drive unit. If you need several drive units so the data can be accessed on several machines, the cost of the additional drive units must be factored in.

---

not the other way around. For example, you can use an 88MB cartridge in a 200MB drive but not in a 44MB drive. To make this a bit more confusing, the original 88MB drives could read from, but not write to, 44MB cartridges. Newer 88MB drives, known as 88c, are now fully compatible with the 44MB cartridges.

While it's not the dominant system it once was, as a class, the $5^1/4$-inch system is the most common in the computer graphics field. Therefore, this system is the most compatible with other sites you'll want to exchange data with, such as service bureaus, designers, and publishers. This is especially true on the Macintosh platform.

Unfortunately, SyQuest cartridges tend to be rather fragile. Even though the system's most sensitive components are in the drive, the mechanism that supports the platter in the cartridge is delicate.

Dropping a SyQuest cartridge from the height of a table top, even in its padded case, will likely result in some lost data, and maybe a dead cartridge as well. Until recently, there was also initially a design flaw in the drives that let dust enter the area where the head reads the data from the platter. This has been corrected in most of the newer drives by adding a protective flap over the opening where the cartridge is inserted.

While the reliability of SyQuests is probably better than their reputation, the cartridges should always be handled with the utmost care, and your data should always be backed up.

Bernoulli

The Bernoulli system is similar to SyQuest, but instead of using a rigid platter, inside the Bernoulli cartridge is a flexible medium, similar to what's found inside a floppy disk. This type of medium makes Bernoulli cartridges more rugged than SyQuests, so it's less likely that you'll lose data if the cartridge is handled roughly.

The Bernoulli system is available with capacities of 45, 90, 150, and 230MB. Like SyQuests, they are upwardly compatible, so a 90MB cartridge will work in a 150MB drive, but not a 45MB drive.

The Bernoulli system is comparable to SyQuests in both cost and performance. While SyQuest is the removable medium of choice on the Macintosh, Bernoulli is more popular among Windows and DOS users.

M/O: Magneto Optical

M/O devices use a hybrid technology for recording data. Data is written using both a laser and a magnetic head, and is read with the magnetic head. M/O devices have been around for several years but are just now starting to become popular because they are less expensive and more stable than either the SyQuest or Bernoulli systems.

The big disadvantage of M/O systems is that they're considerably slower then SyQuests or Bernoullis, despite recent improvements in their performance. This is especially true when writing data, which requires two passes—once for the medium to be heated by the laser, and the other for the magnetic head to actually record the data. Copying large amounts of data to an M/O drive can be a slow and frustrating experience.

There are two primary standards for M/O drives, $3^1/2$ inch and $5^1/4$ inch. The $3^1/2$-inch system is available in capacities of 128 and 230MB. The $5^1/4$-inch system is available in 650MB and 1.3 gigabyte (1,300 megabytes) capacities. Like SyQuest and Bernoulli, M/O cartridges are upwardly compatible. Unlike the other two systems, M/O cartridges can't read both sides of the disk at once. Each side holds half of the disk's capacity, so the cartridge needs to be flipped in order to access the full capacity.

With the $3^1/2$-inch system, the same cartridge will work in a drive from any vendor. The $5^1/4$-inch drives are less standard because initially some vendors developed their own proprietary formats. The $3^1/2$-inch and $5^1/4$-inch systems are the most common but there are some other less common sizes and capacities as well.

Other Types of Storage

The removable storage systems discussed previously all serve the same general purposes; storage that is economical, accessible, and transportable. There are two other storage systems available that serve more specialized needs.

DAT: Digital Audio Tape

DAT, as the name implies, was originally designed to store digital audio. While DAT never really caught on in the consumer audio market, it is an excellent way to inexpensively store huge amounts of digital data. DAT tapes look like smaller versions of audio cassettes, but with a protective cover on the bottom that prevents the tape from being exposed.

DAT is by far the cheapest and most compact method for storing data. Up to 8 gigabytes of data can be stored on a single tape that costs under $25. The disadvantage of DAT is that you can't just copy data to a DAT tape as you would to a hard drive or removable drive; you need to use a utility to transfer the data onto the tape. While transferring data to the tape is relatively straightforward, retrieving the data can be quiet a challenge. In 8 gigabytes of storage, you will likely have thousands of files. Finding the right file in the directory and then finding the right file on the tape can be a long and annoying process. There are utilities that allow you to copy and retrieve data from DAT tapes as if it were like any other storage medium, but the actual reading and writing of data can still be painfully slow.

There are a few different DAT standards that are not compatible with one another, and the higher capacity systems are the least compatible. If you're simply storing data for in-house use, this is not a problem. If, however, you need to exchange data with another site, don't assume that your DAT tape will be readable at the new site.

Another problem with DAT is a side-effect of its main virtue—the ability to store huge amounts of data. Unless you have several gigabytes of free hard disk space, it's difficult to make a back-up copy of your DAT tape. DAT is somewhat more vulnerable to data loss than other media because the magnetism used to store the data can "read through" a tightly wound tape. In addition, having thousands of your images (which may represent months of work) only existing on a single tape is a scary thought.

CD-ROM: Compact Disc Read Only Memory

CD-ROMs are physically identical to audio CDs, but instead of only storing audio, they can be used to store any type of digital data. CD-ROMs have been available as mass produced commercial products for years, but only recently has making your own, "one-off" CDs become affordable. These systems are known as CD-R, for compact disc-recordable.

CD-R systems record data by using a laser to burn pits into the recordable CD medium. Because the data is physically etched into the medium, CD-Rs are non-erasable, and are the only digital medium that does not use magnetism for storing data. As a result, CD-ROM is the most stable method for storing your data. Adding to the security of the data is the fact that the CD itself has no moving parts, making it more rugged than other removable media. CDs are not indestructible however, and like all media, should be handled with care.

Each CD-R disk can hold up to 650MB of data, and can be read by any standard CD-ROM drive, making them very compatible.

Conclusion

File formats and compression can be confusing areas but they don't need to be. The step after scanning in your production process will determine the correct file format to use. You may need to experiment a bit to find out which formats are most suited to your needs. The decision to compress your image should be determined by your quality requirement, the amount of storage you have available, whether you need to transmit images via modem, and whether your printer and the other applications you use can accept compressed files.

As for storage, the first priority is a big, fast hard drive. At some point you'll want to use a removable storage system to reduce the cost of saving images and to make transporting them more convenient. The type of removable system you chose will depend on its cost, performance, and compatibility with other sites.

While it may not be very compatible or convenient, DAT is the most cost effective solution for storing massive amounts of data. For protecting irreplaceable images, CD-R is the best choice.

No matter what type of media you use, remember that any media will eventually fail, so for all your important data, always keep a backup.

C H A P T E R

8

Calibration

What Is Calibration?

Seeing Color

Starting at the End, Again

Approaches to Calibration

The Steps of Calibration

Color-Management Systems

Anyone who's ever worked with color desktop publishing at some point has probably asked: "How come the image on my monitor doesn't look like the original, and the printed image doesn't look like either of them?"

The answer is: There is no reason why these images should look anything alike unless your production system has been properly calibrated. This is because film, scanners, monitors, and printers don't speak the same language when it comes to color. Every device used for color publishing uses a different technology to capture and display color. Not only do they describe color differently, their color gamut—the range of colors each device is capable of representing—is different as well. (See the sidebar, "Converting Out-of-Gamut Colors," at the end of this chapter.)

What Is Calibration?

Calibration is the practice of adjusting some or all of the devices used for displaying and capturing images so that color is handled in a predictable and visibly consistent manner throughout the entire production system. To do this, every step in the production process must be coordinated from the perspective of color.

The goal of calibration is to preserve the color of the original to the best of the production system's ability. This is done by optimizing each device so it will give the most accurate and consistent results it is capable of, and by developing a color relationship between devices.

Seeing Color

Calibration is a considerable challenge not only because color is handled differently by every piece of equipment, which is discussed later in this chapter, but also because seeing color is a very subjective experience. Calibration involves precise measurement and adjustment, but this shouldn't obscure the fact that the goal of all this precision is to control something that is itself imprecise.

Defining Color

Everybody knows color when they see it, but what are we really seeing? Color is made up of selected portions of the visible light spectrum (see Figure 8.1). When all the light in the visible spectrum is present in full and equal amounts, we see white (see Figure 8.2). When no light is present, we see black. When less than the entire visible spectrum (white), but more than none (black) is present, we see a color or a shade of gray.

Figure 8.1
Color is created by light waves, which are part of the electromagnetic spectrum. The light waves that our eyes are sensitive to are known as the *visible spectrum*.

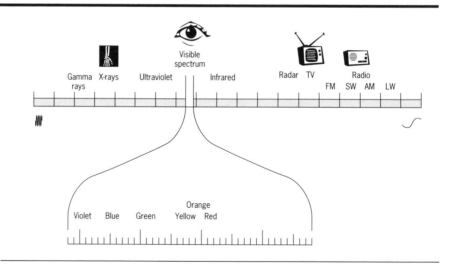

Under white light, an object appears a certain color because it absorbs some colors of the visible light spectrum and reflects others. The color that is reflected is the color we see (see Figure 8.3). For example, an object appears blue because it absorbs all the light in the visible spectrum, except for the waves that cause us to perceive blue, which are reflected.

Figure 8.2
White light is made up of all the colors of the visible spectrum.

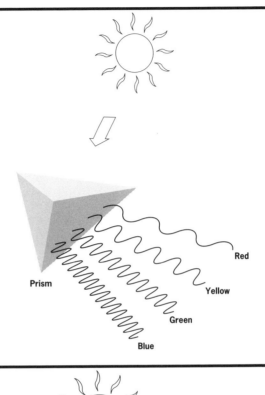

Prism

Red

Yellow

Green

Blue

Figure 8.3
An object appears to be a certain color because it absorbs some light waves and reflects others. The color we see is based on the light waves the object reflects.

This works nicely in theory, but in fact most light sources do not emit all of the color frequencies in equal amounts, but are biased toward one color or another. For example, fluorescent light tends to be biased toward green, and incandescent lights are biased toward yellow.

Seeing with Our Brains

You may wonder why, when viewing a white object under fluorescent light, it doesn't appear green. This is because in our brain, we automatically compensate for the color bias of the light source. Unless a light source is heavily filtered, it appears to us as white.

When light waves reach the back of our eye they excite specialized areas of the retina, which send a signal to the brain. It is in the brain that the sensation of color occurs. While the light waves that cause us to see color are objective and measurable, the actual sensation of color is not.

The fact that color itself can't be measured is a significant challenge to color calibration. As mentioned above, every device in a color publishing system uses a different method to describe color. Film displays colors by varying the densities of dye in the film. Monitors display colors by causing red, green, and blue phosphors to glow at different intensities. Offset printing creates color by controlling the amount of ink applied to paper. The physical methods of measuring color can't be readily transferred from one device to another, but instead must be interpreted.

Memory Colors

Our perception of color is also influenced by a phenomenon known as "memory colors." Here's an example of how memory colors works.

Consider an object, like an apple. An apple looks pretty much the same color to us whether we're looking at it under fluorescent light, incandescent light, or daylight. If we actually measured the frequency of the light waves reflected off the apple under each of these conditions, the readings would be quite different. If the color information we receive is different, why does the apple always appear to be the same color?

The apple always looks the same because our brain automatically adjusts the colors of common objects so they look natural, based on our memory of what those items look like. We know from life experience what certain common, everyday colors should look like, such as a red apple, a blue sky, green grass, and flesh tones. These are known as memory colors. Often our expectation and experience determine the colors we see as much or more than objective factors such as the light source. In fact, the definition of "common objects" depends on many factors. As a result, one's perception of color is also affected by culture and geography.

Starting at the End, Again

Chapter 2 introduced you to the concept of Starting at the End. To briefly recap, Starting at the End means that every step in the production process must be guided by how the image will ultimately be used. In order to obtain the best results, you must first know which device will be used to reproduce the image, and at what size, before scanning and adjusting the image. This principle is especially important when you are calibrating the devices that will be used to capture, display, and reproduce the image.

Calibration, and color reproduction in general, are made more difficult by the fact that some color information is lost in virtually every step of the production process. The detail and range of colors we are capable of seeing are much greater than those that can be captured on film, which in turn is greater than those that can be printed.

Calibration is often thought of in fairly limited terms, such as optimizing a specific device, or calibrating a monitor to match printed output. In fact, the true scope of calibration is much broader, starting when the picture is taken, and continuing until the printed sheet rolls off the press. (See sidebar "Getting the Colors on Film," on the following page). This section uses the scenario of scanning for printing on an offset press. This form of final output has the most steps, is the most challenging, and is probably the most common.

Controlling Your Environment

Because perception of color is very subjective, you need to control and standardize the light and color of your work environment in order to properly calibrate a production system.

The light source affects how we perceive color, so it's important to always use the same lighting conditions when making color decisions. It's best to use a light source specially balanced for viewing color. The 5000 degree kelvin lights give off light that is made up of almost equal parts of all the colors in the visible spectrum. It may not be practical to light an entire work area this way, but at the very least, you should have a color viewing area for evaluating original photographs and printed color.

It's also best if the color of the walls in the room in which you work are a neutral gray. If the walls are bright blue, for example, some of the direct light will reflect off the walls, giving the light in the room a blue bias. Even the light reflecting off your clothes can affect the way you view color, so avoid brightly colored clothes when doing critical color work.

Once you've calibrated your monitor, you should leave your monitor's brightness and contrast settings alone. Otherwise, the contrast range that was carefully set with calibration software will be altered. Like the color of the walls, the desktop (background) color on your monitor should also be set to a neutral gray. The way we see color is affected by nearby colors, and a bright and colorful desktop will make it difficult to evaluate what you see on your monitor.

Approaches to Calibration

The traditional approach to calibration is to carefully control each step, with the goal of optimizing each device and aligning all of the devices with the final output. A newer approach, covered later in this chapter, is to use color-management systems, which more or less automatically optimize each step of the production process to provide consistent color to the best of each device's ability.

Getting the Colors on Film

Have you ever wondered why your snapshots don't look the way the scene did when you photographed it? The problem isn't that the film incorrectly captured the subject, but rather that your brain controlled the way the subject looked to you. Because film has no brain, it can only objectively record the impact of light on the scene.

A professional photographer tries to capture on film what he or she sees, unless the goal is to create a stylized image. Because most light is not evenly balanced, a specific combination of film, filter, and type of processing must be chosen to compensate for the color bias of the light source. For example, fluorescent lights are usually biased toward green. When using film balanced for daylight under a fluorescent light, a magenta filter must be used to compensate for the green cast that would otherwise show up on the image.

By choosing a specific combination of film, filter, and processing (and lighting if the photographer can control it), a photographer can produce a transparency that captures the subject in a way that looks natural.

In-House or Out-of-House?

If all of your production work is done in-house, at least from the scanning to the imagesetting stage, then calibration is a multifaceted, but fairly straightforward affair. If, however, you have some of your work done out-of-house, you need to make sure that your vendor's system is calibrated. The vendor should be able to provide you with printed test targets to which your monitor can be adjusted. To make sure that the color meets your expectations, it's a good idea to run several tests before sending out an important job. It pays to ask your vendors how, and how often, they calibrate their system.

The Steps of Calibration

As stated earlier, the goal of calibration is to get consistent, predictable color from your production system. To do this, each device must be optimized to reproduce color as accurately as possible, and also to create a relationship between all of the devices in regard to color. The following sections examine how each device in the production system is calibrated.

Adjusting the Scanner

Uncalibrated scanners can vary in the way they record color from the same original due to the intensity and color bias of their light source, the sensitivity of the CCDs they use, and the filter set.

Most mid- to high-quality scanners perform a self-calibration. This takes place either when the scanner is first turned on or when

---

What about Video?

This chapter deals mainly with calibration for printed output. But what if your images won't be printed, but will instead only be for displaying on a monitor, such as in a video or multimedia presentation. What can you do to control the color for this type of "output?"

Unfortunately, the answer is: "Not much." Unless your images will be used in an environment you control, such as a kiosk or at a workstation in your facility, there is really nothing you can do to standardize how your images will be displayed. Because monitors are not standardized in the way they display color, you have no way of knowing the characteristics of someone else's monitor. (See the sidebar "All Monitors Are Not the Same," later in this chapter.) This is an even greater problem when producing images that will be widely distributed, such as those on commercial CD-ROMs or posted on line.

One positive note is that colors don't need to be as precisely controlled for display on a monitor as they do in print in order to look natural. This is because monitors provide their own source of light and we tend to perceive even biased light sources as a neutral white. If the white areas of the image appear neutral to us, the rest of the image will look balanced. Of course, this illusion will dissolve if the screen is heavily biased, or if the image on the monitor is compared to the original scene or photograph.

the user instructs it to do so through the scanning software. This self-calibration allows the scanner to adjust its sensitivity to the light source so the entire image area is properly exposed.

When the scanner is first turned on, the light source will vary in intensity. It usually takes about a half hour for the light and the rest of the components to stabilize. For consistent results, it's best to let the scanner warm up before it's used.

Software is available that allows you to calibrate your scanner to a particular output device. The software either comes bundled with the scanner or is available from a third party, and it includes a digital image file of a color test target. The target consists of areas of specific color tints and shades of gray. The target is printed on the output device you wish to calibrate your scanner to.

When the printout of the target is scanned, the calibration software compares the color values sent to the printer to those that actually printed. These values are used to create a look-up table. When scans are made, the color information acquired by the scanner is interpreted by the look-up table and converted so that the color values of the original are the ones that are printed.

This method of scanner calibration has its shortcomings because the quality of the output is limited by the accuracy and consistency of your scanner. However, this is one of the easiest methods for getting your output to resemble, if not precisely match, your original.

Manipulating the Monitor

One of the most frustrating things for newcomers to color desktop publishing is spending a lot of time making an image look good on screen, only to have it look different, and usually worse, when printed. Chapter 6, "Understanding and Using Color," discussed some of the problems with trying to match what is displayed on a monitor to printed output. The following sections elaborate on those points and add a few more.

RGB versus CMYK

Monitors display color using the RGB color model, while printed output uses the CMYK color model. RGB has a different, and wider, range of reproducible colors than does CMYK.

Monitors and printed material also differ in the way they create white and black, because of the different color models they use. Monitors create white by mixing equal amounts of red, green, and blue in full intensity. It's virtually impossible to get the white on a monitor to match a white piece of paper. Likewise, black on a monitor is simply the absence of any added color. The darkest part of the image can't be any darker than your monitor is when it's off. If you look at a monitor when it's off, you'll notice that it's not truly black. Hold up next to the screen something that's truly black, such as a vinyl record, and you'll see the difference. Printed black, however, is created with black ink, which is darker and more neutral than black on a monitor.

Transmissiveness versus Reflectiveness

As explained in Chapter 3, originals may be transmissive or reflective. Unless transparency material is used, printed matter is viewed by means of light that reflects off the print from a light source. Conversely, the image on a monitor is its own source of light, which causes video images to have a more luminous look than printed images. While all color is influenced by the light source and surrounding color, this is more true for printed images than video images. (See the sidebar "What about Video?," earlier in this chapter.)

Phosphors versus Ink

Monitors display images by causing tiny specks of phosphors to glow. Printed images use ink, dye, or toner to display the images. Colors created with phosphors have an iridescent quality that printed color doesn't have, while printed colors can be more subtle than those displayed on monitors.

Judging Color on Your Monitor

As a result of all of these differences, trying to predict exactly what printed output will look like based on what you see on your monitor is impossible. But that doesn't mean it's not possible, or desirable, to make your monitor more accurately reflect how printed output will look. While printed proofs should always be the basis for previewing the color for important print jobs, a calibrated monitor will allow you to adjust your images with a reasonable degree of accuracy based on how the image looks on the monitor. Despite the disadvantages of soft-proofing, (viewing color on your monitor in order to predict what the final output will look like) this is still an important step to follow before you go to the expense of making a color print.

As with scanners, it's important to allow a monitor to warm up for a half hour so it can stabilize in order to get consistent results.

All Monitors Are Not the Same

Not only will an image displayed on a monitor look different from printed output, but it will also likely look different on another monitor. The difference is usually most pronounced with monitors from different manufacturers, but even monitors that are the same make and model will display images differently if the monitors are not calibrated. Furthermore, even the same monitor will eventually change the way it displays an image over time. This is because the intensity with which the phosphors glow will diminish as the phosphors begin to wear out.

Different display cards, or built-in video support, also account for why varying monitors may display the same image differently. These cards, or built-in video, control the intensity of the video signal sent to the monitor. The method the video support uses to convert the digital image data into the analog signal required by the monitor is not necessarily standardized or accurate.

Types of Monitor Calibration

There are two methods for calibrating your monitor, software only and hardware. The method you choose will depend on both your needs and your budget.

Software Only: Using Gamma   Calibrating your monitor using only software is easier and cheaper than hardware calibration, but it's less accurate. Software calibration enables you to individually adjust the intensity of each color (red, green, and blue) on your monitor to control the overall tone and the color in the highlights, midtones, and shadows separately. This type of control is like using the color, brightness, and contrast controls on your television set, but with much more precision.

Photoshop includes an excellent software calibration utility called Gamma. Photoshop for Windows has the same function built in, under a control called Calibrate. These tools enable you to equalize your monitor so it displays gray tones as neutrally as possible with a contrast range that matches your output. This works well if you'll be printing to a calibrated output device.

If your output device isn't calibrated, or you're not sure whether it is or not, you can use these utilities to make your monitor match your output. To do this, simply print an image to the final output device and then compare the printout to the electronic image on the screen. You can then use these utilities to adjust your monitor to look like your printed output. Once your monitor is adjusted to your output, you can be confident that your printed output will be close to what you see on your screen. As you use these tools, remember that you are only changing the way the images are displayed, not the image itself. Regardless of how you calibrate your monitor, the image will print the same.

Hardware Calibration   Software calibration does have some shortcomings. One problem with video displays is that the intensity of light across the entire monitor fluctuates minutely all of the time. Another problem is that the maximum and minimum intensity of each color is not necessarily optimized for matching printed output. To compensate for these monitor problems, hardware calibration must be used.

<table>
<tr><td>

Using Gamma and Calibrate

</td></tr>
</table>

The Gamma utility (or the Calibrate function of Photoshop for Windows) is one of the easiest and least expensive (it's free!) methods for calibrating your monitor.

Gamma is a way of defining tonal range. The first step when using Gamma is to set the Target Gamma. Every monitor has a specific gamma. Most monitors have a gamma of 1.8, although some have a gamma as high as 2.5. The gamma of your monitor is probably listed in the manual it came with. If you don't know your monitor's gamma, you can assume that it's 1.8.

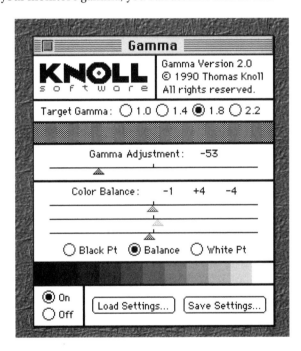

Gamma is a control panel device that is included with the Macintosh version of Photoshop. Using Gamma is a relatively easy and inexpensive way of calibrating your monitor. (The Calibrate controls in Photoshop for Windows provide the same function.)

Now, set the monitor's white point by clicking on the White Pt button. Hold a piece of white paper next to the monitor and move each of the three triangular sliders until the white area on the monitor matches, or at least comes close to matching, the white paper. Match the monitor to the paper stock you'll be printing on if the stock is available.

Next, move the Gamma Adjustment slider until the gray boxes above the controls all have the same brightness. Finally, click the Balance button and move the sliders

Using Gamma and Calibrate

until the strip at the bottom has no color bias and becomes a neutral gray. Do the same thing for the shadow area by clicking on the Black Pt button and adjusting the sliders so the dark areas of the strip are neutral as well. At this point, you may need to go back to the Gamma Adjustment slider and fine-tune the adjustment.

The monitor is now adjusted so that an image on the screen will approximate the way the image will look when printed to a calibrated device. This assumes that your printer has no color bias. You may instead want to make your monitor more closely resemble the output from a particular printer.

To do this, first follow the procedure outlined above. Then print out an image to the printer that you'd like to calibrate your monitor to. The image you choose should include a range of colors and tones. Included with Photoshop is an excellent image to use for calibration, called Olé No Moiré, located in the Separation Sources Folder. After printing out your image, compare the printed image with the way it appears on the monitor. If the colors are substantially different, click on the Balance button in Gamma, and move the sliders until the colors on the monitor match those in the print.

Hardware calibration devices usually consist of a video board, software, and a measuring device that reads the light emitted from your screen. The measuring device reads the light intensity from the screen and creates a look-up table of the values. Once these values are recorded, the image to be displayed is processed through the board so that the intensity of light will reflect the actual image data and better match printed output.

Hardware calibration devices can either be used as add-ons to your existing monitor, or you can buy a monitor specifically designed for prepress work that comes with its own hardware calibration tools. The add-on devices start at less than $1,000, while specially designed monitors can cost several thousands of dollars. Hardware calibration does a much better job of matching your monitor to printed output than does software calibration. However, due to the reasons stated earlier, the differences between the way monitors and printers display color limits how closely they can match, even with hardware calibration.

Color Printers

As mentioned in Chapter 2, "The Production Process," digital printers may either be used for final output or as a proofing method for offset printing. In either case, for best results the printer should be calibrated. Most high-end digital printers (such as the 3M Rainbow Printer and Iris ink jet printer) come with their own calibration software. Others either offer it as an option, or can use calibration software from a third party.

The goal of calibrating a printer is to equalize it, so that a specific tint sent to the printer is the tint that is printed. Equalizing a printer is done with printer calibration software. A test sheet—with varying tints—is printed. If the color printer is used for final output, the printed tints are read with a densitometer or compared with a control test sheet to evaluate the tints that printed. If the color printer is used as a proofing device for offset printing, the printed test target is compared with a press sheet. The differences between the tints specified and the tints that printed are entered into the calibration software and a look-up table is built. When the test sheet is sent again, the values sent are interpreted by the look-up table and adjusted so that the colors specified are the ones that print.

Some digital printers need to be calibrated often because they have problems maintaining consistent color. They may print lighter as they run low on toner, as in the case of color laser printers and color copiers. Or clogged jets in ink jet printers may cause lighter printing.

Imagesetters

Like color printers, imagesetters need to be equalized with calibration software. A test sheet is printed, processed, and the values are read with a densitometer. Any tints that do not match those specified are entered into the software and a look-up table is created.

When an image is sent to the imagesetter, the information is intercepted by the look-up table so that the specified tint is the one that prints.

Aside from tint calibration, the density of film output from an imagesetter must also be measured. If the film is overexposed or processed improperly, the film may not be sufficiently dense, allowing light to come through the solid areas. This will cause printed images to become muddy and lose detail.

While calibration software is used to adjust the imagesetter, the real culprit of inaccurate tints and low densities is often the film and processor. Expired or poorly stored film, improper development time, and insufficient chemical replenishment will all cause tints and densities to fluctuate. Frequent calibration of the imagesetter is important, but proper maintenance of the processor is important as well.

Film-Based Proof

Film-based proofing devices (such as those that make Matchprints Cromalins) and Color Keys are usually set up and calibrated by the manufacturer or dealer. It's a good idea to perform periodic tests to see if the tints they produce are accurate and reflect the way your printed output looks.

Offset Printing

Offset presses are calibrated by the pressman according to the type of press and inks used. When a particular job is run, the pressman controls the amount of ink applied to the paper. The pressman adjusts the amount of ink used so that the sheets coming off the press match the proof you supply as the client.

It's important to note that the pressman will adjust the press in order to get the printed pieces to look like the proof. Make sure the proof you supply represents how you want the printed piece to look. On the other hand, if the film you supply is not capable of reproducing the image as it appears in the proof (as may be the case if a digital proof is used) no matter what the pressman does, the printed piece can't look the way you want it to.

Other factors also affect the color of printed images. Pressmen can't adjust the ink for each individual picture but can only adjust the color in strips several inches wide. If several images are in line with one another, as is common when pages are printed in a signature, the pressman must use the same amount of ink for all pictures. This forces the pressman to make compromises so all of the images look acceptable. If you're scanning images to be used in a publication that carries advertising, it is likely that the ads will be favored on press. As a result, the ink will be adjusted to make the ads look as good as possible. Because the ink cannot be adjusted for each individual image, your picture may end up not matching your proof.

Color-Management Systems

As was discussed earlier in this chapter, a major challenge to calibration is the fact that color is device specific. Each device is capable of producing a different range of colors, and uses its own unique method for reproducing them. The step-by-step approach to calibration can be laborious and still may not be precise enough to allow you to accurately predict and control the color you'll get. What is really needed is a way of producing device independent color, which would allow colors to be accurately and consistently reproduced throughout the entire production chain. Color-management systems were developed as an attempt to make device-independent color a reality. The key components of a color management system are:

- A common color language all devices can understand
- Device profiles that characterize the capabilities of each device

- System-level software that enables colors to be converted based on the information in the device profiles.

The following sections discuss all of these components.

A Common Language

For device independent color to work, there needs to be a language for describing color that all devices can understand. One of the CIE (Commission Internationale de l'Éclairage) color models can be used as such a language. The CIE models are based on color perception, as opposed to RGB, CMYK, and other color models that are rooted in the way in which color is created. By describing color as it is perceived, even devices that use different technologies for defining color can create colors that look the same.

Device Profiles

Using a CIE color model provides the entire production environment with a common language, but a device profile is required to interpret this universal language for each specific piece of equipment. To be used with a color-management system, each device used for capturing or displaying color requires a profile. This profile contains a description of its unique color characteristics, such as contrast range, color gamut, white point, and color balance. These profiles are accessed by the color-management software, which converts the image data into a CIE color model, and then into the color commands specific to the display or output device the image is being sent to.

System-Level Software

The system-level color-management software is responsible for coordinating the conversion of color information from one device

profile, to the CIE color model, and back to another device profile. There are several color-management systems currently available, the two most promising being KCMS (Kodak Color Management System) from Kodak and FotoTune from Agfa. Both of these systems are new, so it's still a little too early to talk about the advantages of one system over the other. Hopefully, color-management system software will become standardized so that device profiles will not have to be specific to the brand of color-management software, but can instead be usable by any system.

As standards are developed for the device profiles and the software that interprets them, more vendors will start to create device profiles for their scanners, monitors and printers, and color-management systems will become more widely used. Though they are in their infancy, color-management systems should have a big impact on desktop scanning in the years to come.

Converting Out-of-Gamut Colors

One problem with reproducing color is that not all devices can display the same range of colors, or color gamut. As a result, devices are often forced to try to display colors that they are not capable of reproducing. In theory, these colors should be pushed to their nearest "legal" equivalent. In practice, these colors are usually converted unpredictably and inaccurately. Color-management systems do an excellent job of consistently converting out-of-gamut colors into reproducible colors. Because a device's profile contains information about its color gamut, a video display or a digital proof—both of which have a wider color gamut than offset printing—can accurately display the colors that will be printed.

Conclusion

We'd all like to be able to reproduce real scenes as we actually see them, or at least accurately reproduce photographic originals. Unfortunately, this is close to impossible because of all the translations an image must go through on its way to print. Compounding the problem is the reality that dots of ink on paper can't capture the range of color, detail, and tone of a scene we actually witness, or even a photographic representation of the scene.

The challenges to color calibration are formidable, from the differing technologies used by each device for reproducing color, to the subjective nature of color itself. Color-management systems promise to bring us closer to a time when printed color will be not only accurate and consistent, but also automatic. How far they will take us in that direction is yet to be seen. Until then, we need to be meticulous in adjusting and measuring our equipment and controlling our environment if we want to reproduce color with any degree of reliability.

APPENDIX

A

Halftones

How Halftoning Works
FM Screening

It would be nice if printing presses could reproduce scenes as we saw them, with an infinite variety of shades and colors. Unfortunately, they are much more limited than that. In fact, they can't vary the shade of the dots they print at all. Presses can't print a dot of 25% or 50% density, they can only print solid dots. How can photographic images be reproduced using only solid dots of ink? The answer is halftoning.

How Halftoning Works

In halftoning, the size of dots is varied to create the illusion of different tones. Halftoning is used because printing presses can't vary the densities of their dots, but they can vary their size. White paper that has half of its area covered with small black dots will appear to have a 50% gray tint, as shown here:

This 50% gray area was not created with gray ink; rather, black ink was used to cover 50% of the tint area.

Halftoning has been used for over a century to reproduce photographic images in print. It is used primarily for offset printing. However, some digital output devices use halftoning as well, such as imagesetters (which are primarily used for preparing film for offset printing), laser printers, thermal wax, solid ink, and some color laser printers.

Traditional Halftoning

In traditional halftoning, the original photograph is recorded onto light-sensitive film through a screen. The screen breaks down the continuous tone original into dots. The size of the dots corresponds to the density of the area of the image the dots represent. Lighter areas are represented by smaller dots and darker areas by larger dots.

Here's an example of traditional halftoning:

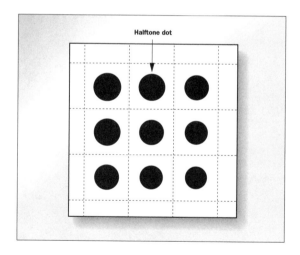

Traditional halftoning creates different shades of gray and color by varying the size of the dots.

Line Screen

The halftone screen is organized into a grid of halftone cells, each of which contains a halftone dot. The number of halftone cells in each linear inch is known as the screen frequency or line screen, and is usually expressed as lpi (lines per inch.) The higher the line screen, the more halftone dots are used to reproduce the image. More dots mean that more detail can be reproduced.

The line screen will usually depend upon the type of press and paper used. The higher the line screen, the more expensive the print job will be because it means printing on a slower press and using high quality paper and inks. See the table below for some typical line screens.

Type of Paper	Line Screen
Newsprint (uncoated)	65–85 lpi
Coated	133–150 lpi
High Quality Coated	150–175 lpi

Digital Halftoning

As discussed above, in traditional halftoning different tones and colors are created by varying the size of the dots. But most digital halftoning devices print all dots at the same size. If you're using an imagesetter with a resolution of 2,400 dpi, every dot is $1/2{,}400$ of an inch. These are known as printer dots.

In order to create the same effect as varying the dot size, these printer dots must be grouped together to make halftone dots. Varying the number of printer dots used to make up each halftone dot has the same effect as varying the size of the halftone dot, as is done in traditional halftoning.

Here is an example of digital halftoning:

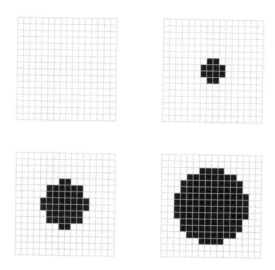

The printer's resolution is the number of printer dots it can create in each linear inch. The line screen is the number of halftone dots per inch. In continuous tone printing, the output resolution is based on the printer's resolution, but in halftone printing, it's based on the line screen.

Resolution, Line Screen, and Shades of Gray

The line screen you use is usually determined by the quality required for your print job. But how do you determine the printing resolution that best suits the line screen you're using? To find the answer, you must first understand how digital halftoning controls the potential number of shades of gray that can be reproduced.

The number of printer dots that make up each halftone dot is inverse to the line screen. For example, let's assume we're printing to a 2,400 dpi imagesetter at 150 line screen. Take the resolution of the printer (2,400) divide it by the line screen (150) to get the number of dots in each linear dimension of the halftone cell (16.) Now square the result to get the total number of dots in the halftone cell (256). In this example, each halftone dot can be made up of as many as 256 printer dots. As a result, the halftone dot can be any of 256 different sizes, thereby creating the visual effect of 256 shades of gray, or color, for each of the process printing colors. The potential to have no dots at all increases the total possible number of values to 257 (see Figure A.1).

Figure A.1
A halftone cell made up of 256 printer dots can be any of 257 values when the potential for no dot is included.

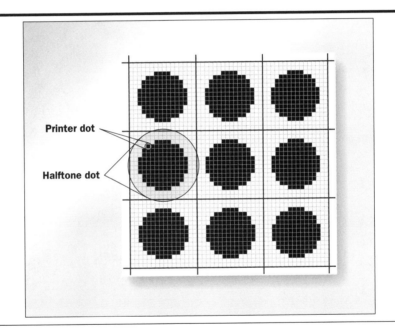

The formula used to determine the number of possible shades of gray divides the printer's resolution by the line screen, squares the result, then adds one. For the example above, the formula looks like this: $(2,400 \div 150)^2 + 1 = 257$.

If your printer does not have a high enough resolution, you'll need to compromise between reproducing detail and reproducing shading. This is because a higher line screen will produce more detail, but at the expense of taking away some dots that make up each halftone dot, thereby decreasing the shades of gray that can be reproduced.

The more shades of gray, or color, the smoother the image will look. Most people can't discern more than 256 shades of any one color. The closer you come to producing 256 shades of each color, the smoother and more realistic your image will appear. The plate "Tonal Reproduction" in the color section of this book shows examples of an image with 256 shades for each color compared to the same image with only 100 shades. (For more on how the shades are combined to make different colors, see Chapter 6, "Understanding and Using Color.")

Screen Angles

Until now, we've only discussed halftoning in the context of one-color printing. But how are the dots arranged when reproducing color images?

Color photographs are reproduced using four colors of ink: cyan, magenta, yellow, and black. If all the colors used the same exact screening, all of the dots would overlap one another, resulting in a muddy mess.

The solution is to rotate the screens of each different color. When the screens are rotated, each color dot is printed at a slightly different angle than the other colors. The number of degrees the screen is rotated is the screen angle. When the four color dots are printed together, the pattern they create is known as a *rosette*, as depicted on the following page.

The screen angles must be very precise in order to get a proper rosette. If the dots are not arranged precisely, the rosette will not be tight enough and the image will look coarse. A bigger problem is that the dot pattern created by the line screen and screen angle may be visible when the image is printed, resulting in what is known as a *moiré pattern*. Moiré patterns usually appear as a plaid-like pattern of criss-crossing diagonal stripes.

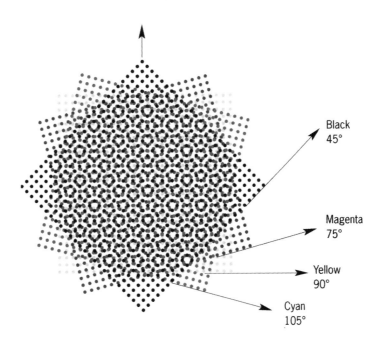

Getting moiré-free screening is tricky business. Realizing this, imagesetter manufacturers have developed their own proprietary screening methods. This screening information is applied as the image is processed in the RIP (raster image processor).

FM Screening

A new type of screening has recently made its commercial debut, and it has the potential of making halftoning obsolete. Unlike halftoning, which uses variable sized dots in fixed locations, FM (frequency modulation) screening uses fixed sized dots, and varies their location. Halftoning makes an area darker by making the dots bigger, FM screening darkens an area by printing more dots.

There are several advantages to using FM screening over halftoning. The dots used by FM screening are much smaller than halftone dots, so this method is capable of reproducing much finer detail than can halftoning.

Problems like moiré patterns or chunky rosettes don't exist with FM screening because unlike halftoning, FM screening doesn't place dots in a regular pattern. In fact, FM screening does away with the whole process of using line screens and screen angles.

Because FM screening is so new, there are still some problems that need to be resolved. One problem is that dot gain, the amount of ink that spreads when the ink hits the paper, is accentuated with FM screening. This is because dot gain occurs around the edges of the dot, and when a lot of small dots and a few larger dots cover the same size area, the small dots have more edge area. Another problem is that the plate material used for reproducing the larger halftone dots is not sensitive enough to properly record the smaller dots used in FM screening.

Most of the problems with FM screening are the result of it being such a new method. On the other hand, because halftoning has been around for over a century, a tremendous body of knowledge has been built around it, along with the required equipment and materials. Despite its advantages, it will take a little while before FM screening becomes as ubiquitous as halftoning.

APPENDIX

B

Digital Cameras

Filmless Photography
How Digital Cameras Work
Choosing a Digital Camera
What about Quality?
Should You Get a Digital Camera?

Scanning turns a photographic original into a digital image. But what if you could eliminate the use of film altogether and create digital images directly from the scene you want to capture? You can do this by using a digital camera.

Digital cameras are similar to conventional cameras, but instead of using film to record the image, the image is recorded by a CCD (charged coupled device). Digital cameras capture a scene in very much the same way a scanner captures the image from a photographic original. In fact, you can think of digital cameras as scanners that record real-life scenes instead of photographic originals.

Filmless Photography

If a photograph is to be scanned and digitized as part of the production process, digital cameras offer some significant advantages over conventional film-based photography. These advantages are realized as instant access to images, lower costs, as well as other benefits.

Instant Access to Images

When a scene is captured using a digital camera, the image is available instantly. There is no need for the usual delay of film processing and scanning. Even under optimal conditions, it can take several hours to first process and then scan a photographic original. Even then, the rush charges may make such quick turnaround unaffordable.

The ability to immediately see the result of a photo shoot, or even a single exposure, gives photographers instantaneous feedback on their work. This enables them to fine-tune their shoots in order to get exactly the image they want. The primary technique used today for previewing the results of photographic images is the Polaroid instant picture. Polaroids are useful for evaluating a pictures composition and lighting, but they still don't give a true representation of how the final image will look.

Lower Costs

Digital cameras can offer a significant cost savings over film-based photography. By eliminating the expense of buying film processing, and paying for scanning, digital cameras make the cost of taking additional photographs marginal. If you plan on doing all of the production work in-house, digital cameras also eliminate the expense of both a scanner and a scanner operator.

Other Benefits

The environment also benefits from the use of digital cameras. Avoiding the use of film eliminates the need for the silver, dyes, and petroleum products used to manufacture film, as well as the resources used and the waste created by its packaging. Digital image capture also avoids the need for the hazardous chemicals used for film processing.

How Digital Cameras Work

The heart of a digital camera is the CCD array. This is the part of the camera that actually records the image and begins the process of turning it into digital information. There are two different types of CCD systems used: linear arrays and area arrays. The type of array the camera uses to a large extent determines the cameras cost, resolution, and how quickly it captures an image.

Linear Arrays

In a linear array, the CCD is made up of photoelectric cells (the elements of the CCD that actually turn the light into electrical signals) aligned in a strip. The number of cells in the strip defines the vertical resolution. When the exposure is made, the CCD moves in tiny steps in order to capture the image. The number of steps defines the horizontal resolution (see Figure B.1).

Figure B.1
A linear array creates and captures an image one column of pixels at a time. It moves in steps to create the entire image.

It can take several seconds or even minutes to make an exposure with a linear array camera because it takes time for the CCD to travel across the image area. As a result, cameras that use linear arrays are suitable only for photographing inanimate objects. The slow exposure also means photographic strobes and flashes won't work with these cameras. The duration of the illumination from these lights is shorter than the time required to make the exposure, so only part of the image will be properly illuminated. Instead, constant light sources, known as "hot lights" must be used.

Linear arrays do have the advantage of offering more resolution for the money than area arrays. This is because linear arrays require significantly smaller CCDs, which contain fewer photoelectric cells than area arrays. Currently, it's too expensive to manufacture area array CDs that can match the resolution of linear arrays.

Area Arrays

Area array cameras use a rectangular shaped CCD. The number of photoelectric cells in each of the horizontal and vertical dimensions defines the cameras resolution (see Figure B.2).

Figure B.2
An area array captures all of the image data at one time by using a matrix of photoelectric cells.

Area array cameras capture the entire image at once, so exposure times can be as fast as $^1/_{2,000}$ of a second, rivaling the shutter speeds of conventional film cameras, making them suitable for most situations. Area array cameras are less economical in terms of resolution because the larger chips make the cameras more expensive than cameras that use linear arrays. However, this doesn't mean area array cameras are always more expensive than linear array cameras. In fact, the least expensive digital cameras use area arrays. But the low price and fast exposures offered by these cameras is offset by their relatively low resolution.

Capturing Color

Cameras not only differ in the type of CCD array they use, they also differ in the way they acquire color information. Digital cameras, like scanners, capture color as RGB information. And like scanners, a digital camera may capture all three color channels at once, or in three separate passes, one for each channel. Some linear array cameras use three linear arrays, and capture the red, green, and blue information simultaneously. Some area array cameras capture each color channel separately, and require three exposures. Digital cameras with the fastest exposure times use an area array to capture the red, green, and blue information simultaneously. This can either be done by using three area arrays—each one capturing a different color channels—or by having the light pass through a matrix of red, green, and blue filters, so the light that hits each photoelectric cell has been filtered to represent one of the RGB colors.

Storing Images

Digital cameras also vary in the way they store the images they acquire. The methods include internal memory chips, removable memory cards, built-in hard drives, and the hard drives of external workstations attached to the camera. Each type of storage has its advantages and disadvantages.

Internal memory is convenient, but once it's full you'll need to copy the images onto another medium to continue to use the camera. Removable memory cards solve the problem of memory filling up; you just pop in a new card. But each card still has a relatively limited capacity, and the cards are quiet expensive. Built-in hard drives can store a much greater number of images than can internal memory, but they require a lot of power to operate and therefore can severely tax the batteries' charge. Using a hard drive in a workstation attached to the camera offers high-volume, economical storage, but being tethered to a workstation limits mobility and certainly poses challenges to location photography.

Other Features

When comparing digital cameras, it's important to consider the important features to look for in all cameras. With any camera, the quality of the optics should be examined closely. Depending on its intended use, factors such as size, weight, and durability may be important as well. Finally, consider other features, such as auto and manual exposure modes, built-in flashes, and the ability to synchronize the camera with separate strobes.

Choosing a Digital Camera

The cost and quality of digital cameras vary widely. There are three broad quality and price categories:

- Consumer models priced at under $1,000

- Mid-range models ranging in price from $7,000 to $10,000

- High-end models starting at $20,000 and topping out at over $50,000

Digital cameras take on many forms. The feature set of low-end models falls somewhere between those found on inexpensive 35mm film cameras and disposable cameras. This includes a fixed-focus lens, auto exposure, and a built-in flash. Some digital cameras have a unique design, others use standard film camera bodies, but use a special back that replaces the film with a CCD array. These cameras offer most of the functionality of conventional cameras, such as several modes for controlling exposure, and the ability to use the wide variety of interchangeable lenses available for conventional cameras. However, the CCD arrays are usually much smaller than the size of the film they replace, so the field of view will be narrower than if the same lens were used with film.

When evaluating digital cameras, there are three primary considerations:

- Price
- Resolution
- Exposure time

Any camera you choose will represent a trade-off between these features.

Price

Pricing varies widely, with high-end cameras costing over 50 times more than the lower-end models. In general, paying more for a digital camera will either get you higher resolution, faster exposure times, or both.

Resolution

Depending on your needs and budget, getting high-resolution images may be your priority. Just as in scanning, when using a digital camera, a sufficient amount of data must be captured in order to meet the requirements for the final output size and resolution.

In general, the low-end cameras only provide enough resolution for displaying images on a monitor, or perhaps using the images very small at a low line screen, such as when printing on newsprint. Mid-priced cameras will generally have sufficient resolution for using at an average size in newspapers, or perhaps at a small size in a web printed publication. Only the high-end models have sufficient resolution to capture images for reproduction in high-quality printing, such as catalogues and books.

Exposure Time

The amount of time it takes to make an exposure is also an important consideration when evaluating digital cameras. A camera that can capture high-resolution images as fast as typical film cameras— $^{1}/_{60}$ of a second or faster—is much more expensive than a camera that requires several seconds to capture an image at the same resolution. Generally speaking, if a digital camera can capture images as quickly as a film-based camera, it will either be very expensive, or produce a relatively low-resolution image.

What about Quality?

The quality of the images created with digital cameras varies greatly, depending on the quality of the cameras and the resolution. However, with the exception of the high-end models, the images produced with digital cameras don't match the quality of those recorded onto photographic originals.

There are two major bottlenecks to getting quality images from digital cameras: resolution and dynamic range. Even a relatively small film format, such as 35mm, can capture more image information than all but the highest-resolution digital cameras. The resolution of an image acquired with a digital camera is dependent upon the size and type of CCD array it uses. But no matter what type of array is used, it's very expensive to make a camera that can capture enough image information for most uses.

Film can also capture a much greater dynamic range than most digital cameras can. Some of the newest and most expensive ($40,000 and up) digital cameras can match or even exceed the tonal range of most film formats. As discussed in Chapter 8, a medium's dynamic range is its ability to reproduce tonal steps in the highlight and shadow area of an image. Capturing the greatest possible tonal range is important if the image will be reproduced as a continuous tone print, or if the same image will be reproduced using several different methods, each with its own unique tonal range. While film can reproduce more tones than digital cameras, the high-end cameras can capture more tonal information than can be reproduced on an offset press.

Digital cameras do have some quality advantages over film-based photography. Each type of film has its own particular color characteristics. Some film is formulated to give emphasis to primary colors, while others may biased toward warmer or cooler tones. Film also suffers from color shifts in the shadow and highlight areas. Digital cameras can record color in a more neutral manner.

Of course there are still many advantages to traditional photography. As mentioned above, photographic film can capture a much wider tonal range than can the CCD in most digital cameras. Film is also a much more cost-effective and compact storage medium for images than is digital storage. Some film is also formulated to capture images more quickly and under lower lighting conditions than most digital cameras can. This enables photographers to shoot at faster shutter speeds to freeze action, and with smaller apertures to increase the field of focus.

And while the marginal cost of using additional photos is much less with a digital camera, conventional cameras are much cheaper than comparable digital cameras.

Should You Get a Digital Camera?

Digital cameras are particularly suited to several photographic situations. Newspaper photographers will benefit from the time savings afforded by digital photography. Not only is the time spent waiting for the film to be processed and scanned eliminated, but if the picture was shot at a remote location, transmitting a digital image via modem is a relatively fast and straight-forward process.

Product catalogue photographers can benefit from the use of high-end digital cameras. High-quality results can be obtained in the controlled lighting of a studio, and long exposure times don't matter with inanimate objects. The savings on film and processing can be enormous.

Even low-end cameras can be useful for publishers of real estate and auto shopper guides. These publications usually reproduce the images at relatively small sizes, print at a low line screen, and are very price sensitive. The savings and convenience of using a digital camera for this type of publication can be very appealing.

Unless you use the highest-end digital cameras, conventional film is still the best method when images will be reproduced at a very large size, when capturing the maximum tonal range is important, or when the final product will be photographic print.

Conclusion

While digital cameras are still a long way from replacing film-based photography, they have emerged in the last few years as a viable alternative for some situations. In some specialized uses, digital cameras have significant advantages over conventional photography. As the quality of digital cameras improves, and prices drop, digital cameras will likely become as common as film cameras are today.

APPENDIX

PhotoCD

What Is PhotoCD?

Putting Your Images on PhotoCD

How PhotoCDs Are Made

Proprietary Format

Maximizing Quality

Other PhotoCD Formats

Kodak PhotoCD has created quite a buzz among imaging professionals. Originally designed as a low-cost way for consumers to view snapshots on their television sets and to make digital prints, PhotoCD has become a tool for professional photographers, publishers, and business users. They are attracted by the low cost of the scans, the accessibility of the images, and the ease with which they can use the service.

What Is PhotoCD?

PhotoCD was designed primarily as a method for storing and displaying photographic images. The images on a PhotoCD can be accessed on any computer platform, and can be viewed on a standard television set equipped with a PhotoCD compatible player. The original PhotoCD standard is known as the Master PhotoCD disc. In recent years, new variations of PhotoCD have been developed to meet different needs. They include Pro PhotoCD, which can handle larger-sized originals and scan at a higher resolution than can the Master PhotoCD system; Print PhotoCD, which can be used to store entire documents, complete with text, graphics, and CMYK images; and Portfolio PhotoCDs, designed for use with interactive presentations, complete with sound and scriptable navigation.

Putting Your Images on PhotoCD

The goal of the original Master PhotoCD is to make digitizing and storing images inexpensive and easy. You simply bring unprocessed 35mm negative or slide film, or already processed slides and negatives to a photo finishing lab. What you get back is a CD-ROM containing scans made from your originals, and an index of the content of the disc in the form of a color print with thumbnails of each image. The cost ranges from $1 to $5 per image. You can then view the images on your television set if you have Kodak's PhotoCD player or Phillips' CD-I (Compact Disc-Interactive) player. More importantly, you can access the images with your personal computer if you have a PhotoCD compatible CD-ROM drive and software that can recognize and retrieve PhotoCD images.

Each CD holds up to 100 images. Like any CD-ROM, once an image is written to a PhotoCD, the image can't be erased or changed and written back to the disc. As long as there's room left on the disc, you can bring it back to the photo lab and have them add more images to the same disc. In order to access these additional images, a multi-session compatible CD-ROM is required. Most CD-ROM drives currently available are both PhotoCD and multi-session compatible.

How PhotoCDs Are Made

PhotoCDs can only be made on Kodak PhotoCD imaging workstations. The price of these workstations starts at $150,000, so unless you use tens of thousands of scans annually, or want to provide PhotoCD services to others, you'll need to send your images out to a PhotoCD service center. Some local photo labs may have their own PhotoCD authoring system; other labs will need to send them to a service center.

The PhotoCD authoring system includes a very fast scanner with an automatic film loader capable of scanning originals in just seconds. The scanner is attached to a UNIX workstation running software that performs color and tonal adjustments to the scans, and converts the image into Kodak's proprietary image pack format. After the images are processed, a CD-writer records the image data onto a writeable CD-ROM. Thumbnails of the images are then output to a color printer to create the index.

Proprietary Format

There's more to PhotoCD than just scanning images and storing them on a CD. What makes PhotoCD unique is the method used for storing the images. In order to improve the portability of the images, and to make them more compact, Kodak developed their own format known as an image pack. The image pack is created by converting the scanned image into a proprietary color model known as PhotoYCC, and then splitting the image data into five interdependent components, each one representing a different resolution.

PhotoYCC

The Kodak PhotoCD imaging workstation converts the RGB data acquired by the scanner to PhotoYCC format. PhotoYCC consists of three channels of image information, one luminance (brightness) and two chrominance (color). Saving the image in this manner allows more flexibility for controlling color when printing to a wide range of devices. Saving the image with separate luminance and chrominance channels also allows for a very efficient compression scheme. This is possible because people notice changes in brightness more than changes in color. Therefore, the chrominance channels can be given more compression than the luminance channel, allowing a high level of compression, while only losing a negligible amount of visual information.

Five Resolutions

After the image is converted to PhotoYCC, it is processed to create five different files, each of which enables you to access the image at a different resolution (see Figure C.1). The original scan at a resolution of 2048 × 3072 pixels is compressed down to what's known as a base image of 512 × 758 pixels. But the data not included in the base image is not discarded. Instead, it is saved as "residual" information, and is added to the base image when a higher-resolution version of the image is acquired. The image pack is created when these five components are put together with a preview into a single file.

Figure C.1
When you acquire an image from a Master PhotoCD, you can choose to open it in any of five different resolutions.

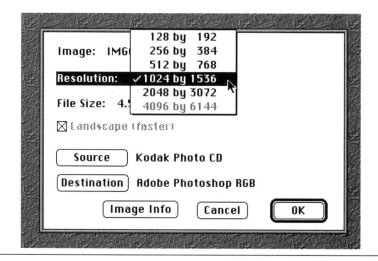

The five different resolutions available are expressed as factors of the base resolution. See the table below.

Image Component	Pixel Dimensions	Size as RGB File
base/16	128 × 192	72K
base/4	256 × 384	288K
base	512 × 768	1.13MB
4base	1024 × 1536	4.5MB
16base	2048 × 3072	18MB

Maximizing Quality

The quality of PhotoCD images depends on two factors: the care taken by the PhotoCD service provider, and the way the images are accessed from the disk. Not all PhotoCDs are created equal, and not all software can access PhotoCD images equally well.

Use a Quality Service Center

The care and time taken by PhotoCD vendors when scanning and processing your images has a significant impact on the quality of the images. A good PhotoCD provider will clean your original before it's scanned. This lets you avoid the hassle of cleaning the same pieces of dust off the image each time you use it, thus saving valuable post-processing time. They will also take the time to set up the scan properly, so it will more closely match the original. A PhotoCD vendor that does high quality work is also more likely to provide good service, such as handling your originals with care and professionally tracking your job. The premium charged by these service centers is usually money well spent.

Use the Right Software

The way images are acquired from the PhotoCD also affects the quality of the image. The best way to acquire the images is to use software that allows you to convert them from PhotoYCC format directly to CMYK. This minimizes the loss of color information when you are making separations. For best results, you should use the KCMS (Kodak Color Management System) when acquiring the images from the disc (see Figure C.2). This will ensure that the color information you acquire is best suited for the type of output device you'll be using. (See the plate, PhotoCD, in the color section of this book.)

Figure C.2
Kodak's
Precision
Color Starter
Kit allows you
to acquire
images from
a PhotoCD
direct to
CMYK
separations.

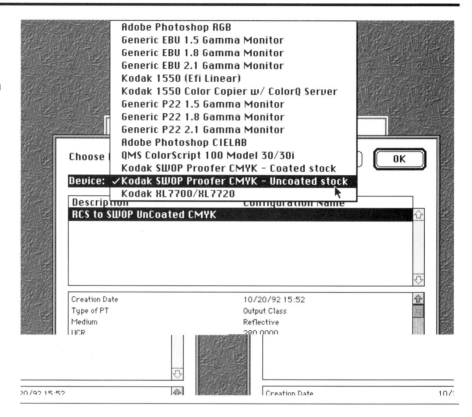

Other PhotoCD Formats

Kodak has taken the core components of the original master
PhotoCD, and created some variations targeted towards specific
types of image storage and display. This is only a partial list of the
PhotoCD variations currently offered, and more are undoubtedly
under development.

Pro PhotoCD

The Pro PhotoCD system is similar to the original PhotoCD format,
but it has several enhancements geared toward professional photographers. Pro PhotoCD imaging workstations are not limited to
35mm originals, but can accept film negatives or transparencies as

large as 4×5 inches. They are capable of scanning at a resolution of 4096×6144, twice that of master PhotoCD. The result is a 64base image, weighing in at a whopping 72MB. The higher resolution scans means that as few as 25 images will fit on each disk.

Pro PhotoCD also offers password protection for access to the high-resolution images, and areas for entering caption and credit information. The scanning and processing of the images for Pro PhotoCDs are given more care, reflecting the needs of the more demanding client base.

Of course this enhanced quality, resolution, and added features comes at a price. Pro Photo CD scans can cost $20 to $25 per image, still a relatively good price for a high-resolution scan.

Print PhotoCD

Print PhotoCD was developed as a method for exchanging prepress data files of all types, including text and graphics as well as images, between high-end electronic prepress systems. Print PhotoCDs can also contain page layout documents and page geometry information, so entire press-ready pages can be stored and transported. The images stored on Print PhotoCDs can be in either PhotoYCC or CMYK format, enabling them to store separations created by high-end scanners. Most of the major electronic prepress vendors such as Scitex, DuPont, Crossfield, and Linotype-Hell have agreed to support Print PhotoCD. Unless you work with these high-end systems you needn't concern yourself with Print PhotoCD.

Portfolio PhotoCD

Portfolio PhotoCD is designed to allow you to create presentations using the PhotoCD format. Portfolio PhotoCDs can store up to 800 images, as well as sounds and other graphics. Most importantly, they allow you to build scripts that are triggered by buttons, allowing you to make interactive presentations.

Conclusion

PhotoCDs can be an attractive alternative to doing your own scanning. Not only are they an inexpensive way to get relatively high-resolution scans, but they also come on their own storage medium.

PhotoCD isn't a replacement for all scanning situations. Unless you use Pro Photo CD, the highest resolution PhotoCD image may not be high enough. Also, you relinquish some of the control you have over image quality that you have when you make your own scans. And unless you have your own PhotoCD imaging workstation, you will get your scans faster by doing your own scanning than by sending them to be put on PhotoCDs.

Still, if you use a quality PhotoCD service provider and access the images using color management system-based software, PhotoCD will likely be able to handle many of your scanning needs. However, if your goal is quality image reproduction, using PhotoCD does not completely eliminate the necessity of understanding color, resolution, and the production process.

APPENDIX

D

Line Art

Challenges of Scanning Line Art

Scanning Resolution

Scanning Tips

Most of this book has dealt specifically with scanning continuous tone originals. However, in many instances, the original you need to scan is not continuous tone, but is instead line art. Line art is made up of solid lines and shapes, with no shades of gray or color. It is typically made up of black lines on white paper, but line art may have colored lines or be on colored paper. Logos, symbols, and pen and ink drawing are examples of line art.

Challenges of Scanning Line Art

Scanning line art poses special challenges. As has already been noted several times, scanning is the process of converting artwork into dots, known as pixels. This works well for continuous tone images such as photographs because each pixel can represent an individual tone. However, turning artwork into dots is a problem when you're starting with solid lines and shapes.

When scanning line art, each sampled area of the original must become either a black or white pixel, the result being a bitmap, or bi-level image. Unless the width of a line is evenly divisible by the size of a pixel, the width of the scanned line will either be wider or narrower than the original line, and may become jagged. (The size of the pixels is based on the scanning resolution. See Chapter 4, "Resolution.") Lines which are not perfectly horizontal or vertical may become jagged because the edges of the line may not sufficiently overlap the photo-sensor in the scanners CCD, and therefore some parts of the line's edge will be represented by white pixels. Curved and diagonal lines are especially prone to becoming jagged. These problems are even more acute when scanning fine lines which may not even be one pixel wide.

Scanning Resolution

In Chapter 4 we discussed how scanning resolution is dependent upon the resolution of the device the image will be printed on. The scanning resolution for line art is also dependent on the printing resolution, but it plays by different rules. When a piece of line art will be reproduced at 100% of its original size, it should be scanned at exactly the printer's resolution, regardless of whether it is going to a halftone or continuous printer (see Figure D.1).

Figure D.1
Both of these images were scanned at 100% of their original size. Note the difference in resolution between the image on the left (100 ppi) and the one on the right (2,500 ppi). The resolution used by the imagesetter on which the film for this book was printed is 2,540 ppi.

For example, when scanning a piece of line art for output at 100% of the original size on a 400 dpi printer, it should be scanned at 400 ppi. If the image will be reproduced at 200% of its original size, then scan it at 800 ppi.

Imagesetters usually print at a resolution of at least 1,200 dpi. Only the highest-end flatbed scanners can scan at a resolution of 1,200 ppi or higher. If you want to reproduce a piece of line art at

100% of its original size to an imagesetter at 1,200 dpi, you can make an enlargement of the artwork on a high-quality copier or have a photographic enlargement made at a reprographics house. If your scanner has a maximum resolution of 400 ppi, you can have your line art enlarged 300%. This will have the same effect as scanning it at 100% of it's original size at 1,200 ppi. ($400 \times 300 = 1,200$.) Make sure that the enlarged version will fit on the bed of your scanner.

Scanning Tips

There are several things you can do, aside from scanning at the correct resolution, to get better results from scanning line art. The first and most important thing you can do to get the best quality results is to start with a good original. Don't expect to get good results scanning images off thermal fax paper or a photocopy which has been through several generations.

Once you have the best quality original possible, the scanning software may allow you to minimize jagged or broken lines. Most scanning software gives you the ability to control the brightness of the scan. If the brightness setting is lowered, more of the edge will be detected as being black, and therefore the lines will be more solid.

Some scanning software allows you to control the threshold when scanning line art. The threshold is the point at which a pixel is considered to be either black or white. If the threshold is set at 50%, an area which is darker than 50% gray will be represented by a black pixel. If the area is lighter than 50% gray, a white pixel is created. If the threshold point is raised, more pixels are considered to be black. If you've scanned the image at the right resolution and you still get broken lines, then raising the threshold may allow you to capture fuller lines with better edge definition.

Most image editing software also gives you control over threshold. One strategy you can use is to scan the line art as grayscale, and adjust the threshold in the image editing software until the lines appear solid. If you do scan line art as grayscale, don't exceed the scanner's true optical resolution. Interpolated resolution will distort the lines. After the scan is made, applying unsharp masking several times will enhance the edge definition of the lines. Remember to scan at the printing resolution and be prepared to deal with a very large grayscale file before you convert it to line art.

One way to get better results when scanning line art with a lot of fine lines is to put a piece of clear acetate between the piece of line art and the scanner bed. This has the effect of magnifying the lines a bit. As a result, the lines will appear thicker in the scan, but at least they won't be broken up.

Conclusion

It would seem that scanning line art would be fairly simple. There's no tonal range, color correction, or calibration to worry about. But scanning line art presents its own unique challenges. You can avoid, or at least minimize, the problems of broken and jaggy lines by starting with a good original, scanning at the printer's resolution, and using the software's brightness and threshold settings. This may not solve all of your line art problems, but it will start you off in the right direction.

E

Vendor Directory

Scanners

Third-Party Scanning Software

Image Editing and Separation Software

Color Printers

Color Management and Calibration Products

PhotoCD Acquire Software

Scanners

Flatbed Scanners

Agfa
100 Challenger Rd.
Ridgefield Park, NJ 07660
201-440-2500
800-685-4271
201-440-8187 fax
Products: Arcus Plus, Horizon Plus,
StudioScan

Apple Computer, Inc.
1 Infinity Loop
Cupertino, CA 95014
408-996-1010
800-776-2333
Products: Apple OneScanner, Apple
Color OneScanner

AVR Technology, Inc.
71 E. Daggart Dr.
San Jose, CA 95134
408-434-1115
800-544-6243
408-434-0968 fax
Products: AVR 3000/GS Plus, AVR
8800/GSX, AVR 6600/CLX, AVR
8800/CLX, AVR 8800T

Computer Friends, Inc.
14250 NW Science Park Dr.
Portland, OR 97229
503-626-2291
503-643-5379 fax
Product: ArtiScan 6000

DPI Electronic Image Systems
69 Old State Route 74, Ste. 1
Cincinnati, OH 45244
513-528-8668
800-374-7401
513-528-8670 fax
Products: Art-Getter 24-Bit Plus, Art-
Getter 30 Bit, Art-Getter R1-2412, Art-
Getter R1-9624, Art-Getter R3-2400

Epson America, Inc.
20770 Madrona Ave.
Torrance, CA 90503
310-289-0770
800-289-3776
310-782-4248 fax
Products: ES-1200C Pro, ES-800C

Hewlett-Packard Co.
P.O. Box 58059, MS #511L-SJ
Santa Clara, CA 95051
800-851-1170
Products: HP ScanJet IIcx, HP Scanjet IIp

Howtek, Inc.
21 Park Ave.
Hudson, NH 03051
603-882-5200
603-880-3043 fax
Products: Personal Color Scanner,
Scanmaster 3+

LaCie, Ltd.
8700 SW Creekside Pl.
Beaverton, OR 97005
503-520-5000
800-999-0143
503-520-9100 fax
Product: Silverscan

Microtek Lab, Inc.
3715 Doolittle Dr.
Redondo Beach, CA 90278
310-297-5000
800-654-4160
310-297-5050 fax
Products: MS-II, ScanMaker II,
ScanMaker IIG, ScanMaker IISP,
ScanMaker IIXE, ScanMaker IIHR,
ScanMaker III

Mustek, Inc.
15225 Alton Pkwy.
Irvine, CA 92718
714-453-0110
800-468-7835

714-453-0101 fax
Product: MFS-6000CS

Nikon Electronic Imaging
1300 Walt Whitman Rd.
Melville, NY 11747
516-547-4355
516-457-0305 fax
Product: ScanTouch

PixelCraft
1300 Doolittle Dr. #19
San Leandro, CA 94577
510-562-2480
800-933-0330
510-562-6541 fax
Product: Pro Imager 7650C

Polaroid Electronic Imaging Systems
565 Technology Square
Cambridge, MA 02139
800-225-1618
617-577-3074 fax
Product: CS-500

Relisys
320 S. Milpitas Blvd.
Milpitas, CA 95035
408-945-9000
408-945-0587 fax
Products: VM4520, VM 3510, VM3410

Scitex America Corp.
8 Oak Park Drive
Bedford, MA 01730
617-275-5150
617-275-3430 fax
Products: Smart 340, Smart Plus PS

Sharp Electronics
Sharp Plaza, Box F
Mahwah, NJ 07430
201-529-9594
Products: JX-320, JX-450, JX-610

UMAX Technologies
3353 Gateway Blvd.
Fremont, CA 94538
510-651-8883
800-652-0310
510-651-8834 fax
Products: UC630 LE, UC840, UMAX
PowerLook

35mm Slide Scanners

Eastman Kodak Digital &
Applied Imaging
343 State St.
Rochester, NY 14650
716-724-4000
800-752-6567
716-724-9261 fax
Product: RFS 2035 Plus

Leaf Systems
250 Turnpike Rd.
Southboro, MA 01772
508-460-8300
800-685-9462
508-460-8304 fax
Product: Leaf 35

Microtek Lab, Inc.
3715 Doolittle Dr.
Redondo Beach, CA 90278
310-297-5000
800-654-4160
310-297-5050 fax
Product: ScanMaker 35t

Nikon Electronic Imaging
1300 Walt Whitman Rd.
Melville, NY 11747
516-547-4355
516-457-0305 fax
Products: Coolscan, LS3510AF

Polaroid Electronic Imaging Systems
565 Technology Square
Cambridge, MA 02139
800-225-1618
617-577-3074 fax
Product: SprintScan

Transparency Scanners

Dicomed, Inc.
12270 Nicollet Ave.
Burnsville, MN 55337
612-895-3000
800-888-7979
612-895-3258 fax
Product: Dicomed Desktop Scanner

Leaf Systems
250 Turnpike Rd.
Southboro, MA 01772
508-460-8300
800-685-9462
508-460-8304 fax
Product: Leaf 45

Microtek Lab, Inc.
3715 Doolittle Dr.
Redondo Beach, CA 90278
310-297-5000
800-654-4160
310-297-5050 fax
Product: ScanMaker 45t

PixelCraft
1300 Doolittle Dr. #19
San Leandro, CA 94577
510-562-2480
800-933-0330
510-562-6541 fax
Product: Pro Imager 4520 RS

Desktop Drum Scanners

Howtek, Inc.
21 Park Ave.
Hudson, NH 03051
603-882-5200
603-880-3043 fax
Product: Scanmaster D4000

Optronics
7 Stuart Rd.
Chelmsford, MA 01824
508-256-4511
508-256-1872 fax
Products: ColorGetter II Prima,
ColorGetter II i, ColorGetter II Pro

ScanView
330 Hatch Drive, Ste. A
Foster City, CA 94404
415-378-6360
415-378-6368 fax
Products: ScanMate 4000, ScanMate
5000, ScanMate Plus, ScanMate Plus II

Screen (USA)
5110 Tollview Dr.
Rolling Meadows, IL 60008
708-870-7400
708-870-0149 fax
Products: DT-S1015AI, DT-S1030AI,
DT-S1045AI

Third-Party Scanning Software

Canto Software
800 Duboce Ave.
San Francisco, CA 94117
415-431-6871
800-332-2686
415-861-6827 fax
Products: Cirrus PowerLite 2.1,
Cirrus PowerPro 2.1

DPI Electronic Image Systems
69 Old State Route 74, Ste. 1
Cincinnati, OH 45244
513-528-8668
800-374-7401
513-528-8670 fax
Product: Art-Scan Professional

Light Source
17 E. Sir Francis Drake Blvd.
Larkspur, CA 94939
415-925-4242
800-231-7226
415-461-8011 fax
Product: Ofoto Version 2

Second Glance Software
25381-G Alicia Pkwy.
Laguna Hills, CA 92653
714-855-2331

714-586-0930 fax
Product: ScanTastic ps

Image Editing and Separation Software

Adobe Systems
1585 Charleston Rd.
P.O. Box 7900
Mountain View, CA 94039
415-961-4400
800-833-6687
415-961-3796 fax
Product: Photoshop

Apple Computer, Inc.
1 Infinity Loop
Cupertino, CA 95014
408-996-1010
800-776-2333
Product: PhotoFlash

Caere Corp.
100 Cooper Ct.
Los Gatos, CA 95030
408-395-7000
800-535-7226
408-354-2743 fax
Product: Image Assistant

Candela, Ltd.
1676 E. Cliff Rd.
Burnsville, MN 55337
612-894-8890
612-894-8840 fax
Product: 4-Seps

Electronics for Imaging
2855 Campus Dr.
San Mateo, CA 94403
415-286-8600
800-285-4565
415-286-8686 fax
Product: Cachet Color Editor

Fractal Design Corp.
335 Spreckels Dr.
Aptos, CA 95003
408-688-8800
800-297-2665
408-688-8836 fax
Product: Color Studio With Shapes

The Human Software Co.
P.O. Box 2280
Saratoga, CA 95070
408-741-5101
408-741-5102 fax
Product: ColorExtreme

PixelCraft
1300 Doolittle Dr. #19
San Leandro, CA 94577
510-562-2480
800-933-0330
510-562-6541 fax
Product: Color Access 1.4

Pre-Press Technologies, Inc.
2443 Impala Dr.
Carlsbad, CA 92008
619-931-2695
619-931-2698 fax
Product: SpectraPrint Pro

ScanView
330 Hatch Drive, Ste. A
Foster City, CA 94404
415-378-6360
415-378-6368 fax
Product: ColorQuartet 3.0

Technosystems USA
3400 Randy Lane
Chula Vista, CA 91910
619-427-0108
800-417-0108
619-427-2180 fax
Product: Chagall

Color Printers

ColorAge
900 Middlesex Turnpike, Bldg. 8
Billerica, MA 01821
508-667-8585
800-437-3336
508-667-8821 fax
Products: ColorQ 1000, ColorQ 2000

Dataproducts Corp.
6219 De Soto Ave.
P.O. Box 746
Woodland Hills, CA 91365
818-887-8000
800-445-6190
818-887-4789 fax
Product: Jolt PSe

Eastman Kodak Digital &
Applied Imaging
343 State St.
Rochester, NY 14650
716-724-4000
800-752-6567
716-724-9261 fax
Products: 450GL, XLS 8300, XLT 7720

Electronics For Imaging
2855 Campus Dr.
San Mateo, CA 94403
415-286-8600
800-285-4565

415-286-8686 fax
Product: Fiery Color Server

FARGO Electronics, Inc.
7901 Flying Cloud Dr.
Eden Prairie, MN 55344
612-941-9470
800-327-4622
612-941-7836 fax
Products: Primera Color Printer,
PrimeraPro Color Printer

GCC Technologies, Inc.
209 Burlington Rd.
Bedford, MA 01730
617-275-5800
800-422-7777
617-275-1115 fax
Product: ColorTone

Hewlett-Packard Co.
P.O. Box 58059, MS #511L-SJ
Santa Clara, CA 95051
800-851-1170
Products: HP DesignJet 650C, HP
PaintJet XL300

Mitsubishi Electronics America, Inc.
800 Cottontail Lane
Somerset, NJ 08873
908-563-9889
800-733-8439
908-563-0713 fax

Products: S3600-30U, S6600-30U,
Shinko CHC-S446i ColorStream/DS,
Shinko CHC-S46i ColorStream/DS Plus,
Shinko CHC-S446i ColorStream/Plus

Nikon Electronic Imaging
1300 Walt Whitman Rd.
Melville, NY 11747
516-547-4355
516-457-0305 fax
Product: Coolprint

Océ-Bruning
1800 Bruning Dr. West
Itasca, IL 60143
708-351-2900
708-351-7549 fax
Product: G5242-PS

QMS, Inc.
One Magnum Pass
Mobile, AL 36618
205-633-4300
800-523-2696
205-633-4866 fax
Products: ColorScript 210, Color Script
230, ColorScript Laser 1000, Magicolor
Laser Printer

Scitex America Corp.
8 Oak Park Drive
Bedford, MA 01730

617-275-5150
617-275-3430 fax
Products: IRIS 3024, IRIS 3047, IRIS
SmartJet 4012

Screen (USA)
5110 Tollview Dr.
Rolling Meadows, IL 60008
708-870-7400
708-870-0149 fax
Product: FP 600S Dye Sublimation
Color Printer

Seiko Instruments USA, Inc.
1130 Ringwood Ct.
San Jose, CA 95131
408-922-5900
800-533-5312
408-922-5835 fax
Products: ColorPoint PSN Models 4 and
14, ColorPoint PSX Models 4 and 14

Tektronix, Inc.
P.O. Box 1000, Mail Station 63-630
Wilsonville, OR 97070
503-682-7377
800-835-6100
503-682-7450 fax
Products: Phaser 220e, Phaser 220i,
Phaser 300i, Phaser 440, Phaser 480,
Phaser IIIPXi, Phaser IISDX

3M Printing & Publishing Division
3M Center, Bldg. 223-2N-01
St. Paul, MN 55155
612-733-4041
612-726-5613 fax
Product: 3M Rainbow Proofing System

Color Management and Calibration Products

Agfa
100 Challenger Rd.
Ridgefield Park, NJ 07660
201-440-2500
800-685-4271
201-440-8187 fax
Product: FotoTune

Candela, Ltd.
1676 E. Cliff Rd.
Burnsville, MN 55337
612-894-8890
612-894-8840 fax
Product: Candela Color Management
System (CCMS)

Daystar Digital, Inc.
5556 Atlanta Hwy.
Flowery Branch, GA 30542
404-967-2077
800-532-6567

404-967-3018 fax
Products: Color Match, Colorimeter 24

Eastman Kodak Digital &
Applied Imaging
343 State St.
Rochester, NY 14650
716-724-4000
800-752-6567
716-724-9261 fax
Product: Precision Input Color
Characterization Software

KEPS, Inc.
164 Lexington Rd.
Billerica, MA 01821
508-667-5550
800-345-6649
508-670-6552
Products: PCS100 Kit, Professional
Color Server XL

Light Source
17 E. Sir Francis Drake Blvd.
Larkspur, CA 94939
415-925-4242
800-231-7226
415-461-8011 fax
Product: Colortron

Technical Publishing Services, Inc.
739 Bryant St.
San Francisco, CA 94107
415-512-1230
415-512-1232 fax
Product: Color Calibration Software

PhotoCD Acquire Software

Candela, Ltd.
1676 E. Cliff Rd.
Burnsville, MN 55337
612-894-8890
612-894-8840 fax
Products: Gamut CD

Eastman Kodak Digital &
Applied Imaging
343 State St.
Rochester, NY 14650
716-724-4000
800-752-6567
716-724-9261 fax
Products: PhotoCD Access Plus,
Precision Device Color Profile Starter
Pack, PhotoEdge

The Human Software Co.
P.O. Box 2280
Saratoga, CA 95070
408 741-5101
408 741-5102 fax
Product: CD-Q Acquire Plug-In

Purup A/S
Sonderskovvej 5
DK-8520 Lystrup, Denmark
45-86-22-25-22
45-86-22-25-11 fax
Product: PhotoImpress

INDEX

A

acetate, use in scans, 172
additive colors, 81
angles, screen, 142–143
area array cameras, 150–151
arrays, CCD, 148–151, 153
ASCII encoding, 97–98

B

Bernoulli system, 109, 110
binary encoding, 97–98
bit depth
    color images and, 77–78, 79–80
    exceeding eight bits, advantages of, 78–80
    grayscale images and, 76–77
    one-bit images and, 75–76
    overview, 75
    pixels and, 43
    shadows and, 79
black, in separations, 81, 83–85
black-and-white originals, 29–30. *See also*
    originals
blankets, 20
blower brushes, 58
booths, viewing, 57

C

calibration
    approaches to, 120–121
    color and, 115–119
    color-management systems in, 131–133
    color printers, 129
    environmental considerations, 120
    film-based proofing devices, 130
    imagesetters, 129–130
    monitors. *See* monitors, calibrating
    offset printing, 130–131

overview, 115
scanners, 122–123
ultimate use of image and, 119–120
for video output, 122
cameras, digital. *See* digital cameras
cartridges, disk, 107–110
CCDs
    in digital cameras, 147, 148–151, 153
    in scanners, 23–24
CD
    CD-ROM, storing images on, 112–113
    Kodak PhotoCD. *See* PhotoCD
channels, 78, 82
charged coupled devices. *See* CCDs
Chromacheck proofs, 19
Chromalin proofs, 19, 130
CIE color models, 132–133
clipping paths, 98–99, 100
CMYK, 10, 81, 124. *See also* color; separations
color
    additive vs. subtractive, 81
    adjustment, 65–67, 69, 70
    bit depth and, 77–78, 79–80
    calibration and, 115–119. *See also* calibration
    CMYK, 10, 81, 124. *See also* separations
    device limitations and, 133
    digital cameras and, 151
    fidelity of, 21
    halftones and, 142–143
    limits of, 78
    in originals, 29
    perception of, 115–119, 120, 121
    in PhotoCD system, 161
    pixel value and, 42
    proofing, 17–18, 19
    ranges of, 86–87
    registration, 32
    RGB, 10, 81, 124. *See also* separations
    scanner software and, 36–37
    scanning in, process explained, 78
    separating. *See* separations
    software for managing, 131–133, 163–164
    standards for, 132–133
color casts, 62, 63, 79–80

color copiers, 11
color gamut, 86
Color Key proofs, 19, 130
color-management systems, 131–133, 163–164
color printers, 10–11, 129
color viewing booths, 57
compressing files. *See* file compression
continuous tone printing, resolution and, 47–48
contrast, adjusting, 65–67, 69, 70
copiers, color, 11
cost
    of digital cameras, 148, 152, 153
    of do-it-yourself scanning, 2, 4, 6
    of mistake correction, 20–21
    of monitor calibration, 128
    of scanners, 23, 24, 26–27
    of software, 38
    of storage, 105, 109, 112
cropping, 36, 65, 68

D

DAT, storing images on, 111–112
DCS files, 96–97
desktop color separation, EPS format and, 96–97
desktop drum scanners, 26–28, 35–36
device profiles, 132, 133
devices, calibrating. *See* calibration
digital audio tape, storing images on, 111–112
digital cameras
    assessing need for, 156
    choosing, factors in, 152–154, 156
    image quality, 154–155
    operation, described, 148–152
    overview and benefits, 147–148
    where to buy, 156
digital color proofing, 17–18
digital halftones, 139–140
digital printers, 10–11, 129
Direct-to-Plate printing, 19
Direct-to-Press printing, 19

disks
    compressing, 102–103
    storing images on. *See* storing images
drum scanners, 26–28, 35–36. *See also* scanners
dust, 56, 58, 69
dye-diffusion printers, 11, 18
dynamic range, digital cameras and, 155

E

emulsion layer, 59
encoding, EPS format and, 97–98
environmental considerations, 56
EPS (encapsulate PostScript) format. *See also* file formats
    clipping paths and, 98–99, 100
    desktop color separation and, 96–97
    encoding, 97–98
    halftone screens and, 99
    overview, 93
    preview options, 94–96
expense. *See* cost
exposure
    evaluating, 61–62
    poor, compensating for, 80
    time of, in digital cameras, 154

F

file compression
    disk compression, 102–103
    JPEG compression, 104–106
    lossless compression, 102–103
    lossy compression, 103
    LZW compression, 92, 93, 103
    modem transmission and, 14
    overview, 102
file formats
    choosing between TIFF and EPS, 98–100
    clipping paths, 98–99, 100
    determining, 14, 15

EPS. *See* EPS format
　　overview, 91, 92
　　PCX format, 101
　　Photoshop format, 101
　　PICT, 100
　　screening information, 99–100
　　TIFF, 92–93
files. *See also* file compression; file formats
　　complexity of, 2
　　size, resolution and, 48–49, 51
film. *See also* photographs
　　for digital cameras, 155
　　film-based proofs, 13, 18, 19, 130
　　printing to, 18–19
fingerprints, 60
flatbed scanners, 23–24
FM screening, 143–144
focus, 33, 64
FotoTune (color-management system), 133

G

Gamma, 126, 127–128
gang-scanning, 35
gray component replacement (GCR), 84
grayscale
　　bit depth and, 76–77
　　in color scanning, 78
　　considering in scanner purchase, 29–30
　　handheld scanners and, 25
　　in threshold adjustment, 172
gray shades, halftones and, 141–142

H

halftones
　　digital, 139–140
　　FM screening, 143–144
　　gray shades, 141–142
　　overview, 137–138
　　resolution, 46–47, 140–141

　　screen angles, 142–143
　　screens, EPS format and, 99
　　traditional, 138–139
handheld scanners, 25
hard drives, storing images on, 106–107
hardware, calibrating. *See* calibration
heat, damage from, 56
high-end drum scanners, 27–28. *See also*
　　scanners
high-key photos, 62
highlights, 31, 61

I

image pack, 161
images. *See also* originals; photographs;
　　　　scanning procedure
　　color in. *See* color
　　importing, 16–17
　　size, consideration in scanner purchase, 30
　　storing. *See* storing images
imagesetters, 18–19, 129–130
importing images, 16–17
ink
　　calibration and, 124
　　separations and, 85–86, 86–87
inkjet printers, 11
interpolated resolution, 50

J

jagged lines, 169, 171
JPEG compression, 98, 104–106

K

Kelvin scale, 57
Kodak Approval System, 18

Kodak Color Management System (KCMS), 133, 163

Kodak PhotoCD. *See* PhotoCD

L

laminated proofs, 19

laser printers, 11

light tables, 57

linear array cameras, 149–150

line art. *See also* images; originals
difficulty of scanning, 169
overview, 29–30
scanning resolution and, 170–171
scanning tips for, 171–172

line screens, in halftones, 139, 140–141. *See also* halftones

literature, in scanner purchase, 34

logos, scanning, 25

lossless compression, 102–103

lossy compression, 103

loupes, 57

low-key photos, 62

LZW compression, 92, 93, 103

M

magneto/optical (M/O) devices, 109, 110–111

Matchprint proofs, 19, 130

memory, image storage and. *See* file compression; storing images

"memory" colors, 118–119

moiré patterns, 142–143

monitors, calibrating
brightness and contrast settings and, 120
color and, 67, 124, 125
differences between monitors, 125
hardware, 126–128
overview, 123
phosphors and, 124

software, 126, 127–128
transmissiveness vs. reflectiveness, 124

N

needs evaluation
originals, types of, 29–30
overview, 28
quality, 30–34. *See also* resolution
speed and productivity, 34–36

negative film, 29

noise, 32

non-emulsion layer, 59

non-laminated proofs, 19

O

OCR, 25

offset printing, 10, 130–131

one-bit images, 75–76

optical resolution, 49–50

originals. *See also* images; photographs; scanning procedure
considering, in scanner purchase, 28–30
cropping, 65
evaluating content of, 61–63
handling, 60
placing in scanner, 63–64
poorly exposed, bit depth and, 80
pre-scan inspection and cleaning, 57, 58–60

output resolution, 46

output types, 10–11

overlay proofs, 19

P

PCX format, 101

phosphors, calibration and, 124

PhotoCD
   cost, 160
   data exchange and, 165
   formats for, 161–162, 164–166
   overview, 159–160
   processing, 163
   for professional use, 164–165
   software for, 163–164
   technology described, 160–161
   viewing options, 160
photographs. *See also* images; originals;
     scanning procedure
   digital cameras. *See* digital cameras
   high-key vs. normal vs. low-key, 61–62
   Kodak PhotoCD. *See* PhotoCD
   traditional vs. digital, 155
photo-multiplier tubes, 26
Photoshop, 38
Photoshop format, 101
PhotoStyler, 38
PhotoYCC format, 161
PICT file format, 100
Picture Publisher, 38
pixels, 41–43, 169. *See also* resolution
placement, scanner, 55–56
plates, 20
PMT tubes, 26
Portfolio PhotoCD, 159, 165–166
positive film, 29
post-processing, 35
PostScript file format. *See* EPS format
prescanning, 15, 35, 36, 64–65
presentations, PhotoCD in, 165–166
PressMatch proofs, 19
printers
   color, 10–11, 129
   digital, 10–11, 129
   halftones and, 140–142
   printer resolution, 46–48
   raster image processors in, 99
printing
   to film, 18–19
   JPEG compression and, 105

   presses, 10, 20, 130-131
   production process. *See* production process
Print PhotoCD, 159, 165
print production. *See* production process
production process
   integrating images, 16–17
   overview, 11–13
   plates and printing, 20
   printing to film, 18–19
   proofing, 17–18, 19
   scanning, 15. *See also* scanning
   separations, 16. *See also* separations
proofing, 11, 17–18, 19
Pro PhotoCD, 159, 164–165

R

RAM, image storage and. *See* file compression;
    storing images
raster image processors, 99
real resolution, 49–50
reflectiveness, 29, 124
registration, 32
removable storage, 107–111
repeatability, in choosing scanner, 32–33
resolution
   in choosing scanner, 33–34
   digital cameras and, 153–154, 154–155
   file size and, 49–50
   halftones and, 140–141
   PhotoCD and, 161–162
   pixels in, 41–43
   real vs. interpolated, 49–50
   scanning resolution, 43–46, 51, 170–171
   scan quality and, 45
   selecting type, 51–52
   setting, 48, 49, 65, 66
   types, 43, 46–48
reusing scans, 17
RGB, 16, 81, 124. *See also* color; separations
RIPs, 99

S

saving scans, 68, 71–72, 91, 92. *See also* file
    compression; file formats; storing images
scanners. *See also* scanning
    calibrating, 122–123. *See also* calibration
    choosing, considerations in. *See* needs
        evaluation
    main types, 23–28
scanning. *See also specific subject*
    basic procedure, 15
    benefits of do–it–yourself, 3–5, 6
    challenges of do–it–yourself, 2–3
    context and, 9–10
    placement, 55–56
    prescanning considerations, 15, 35, 36
    procedure. *See* scanning procedure
    several images simultaneously, 35
scanning procedure
    beginning the scan, 68
    color and contrast adjustment, 65–67, 69, 70
    cropping, 65, 68
    dust, scratches, and fingerprints, 57, 58–60,
        69
    evaluating image content, 61–63
    focusing, 64
    placing image, 63–64
    prescanning, 15, 35, 36, 64–65
    saving, 68, 71–72
    separation, 71
    setting resolution, 48, 49, 65, 66. *See also*
        resolution
    unsharp masking, 69, 71
scanning resolution, 43–46, 51, 170–171
scans, reusing, 17
scratches, 58–59, 69
screens
    halftone, 139, 140–143. *See also* halftones
    RIPs and, 99
    video, calibrating. *See* monitors, calibrating
separations
    automatic, 35
    creating, 83–87
    overview, 16
    RGB and CMYK in, 81

in saving images, 96–97
    in scan process, 71
    viewing, 82
shadows, 31–32, 61, 79
sharpness, of original, 62
sheetfed presses, 10
size
    of images, 14, 16–17, 30
    of pixels, 42
    of preview window, 65
slides, transparency option scanners and, 26
software
    automatic resolution setting via, 48, 49
    bundled with scanner, 38
    controlling threshold with, 172
    interpolated resolution and, 50
    overview and requirements, 36–38
    prescanning and, 64–65
    removing scratches with, 59
    speed and, 35
    third-party, 38–39
solid wax printers, 11
speckles, 32
speed, in choosing scanner, 34–36
storing images. *See also* file compression; file
        formats
    backups, 113
    on CD-ROM, 112–113
    on digital audio tape, 111–112
    digital cameras and, 151–152
    on hard drives, 106–107
    overview, 106
    removable storage, 107–111
    virtual memory and, 108
subtractive colors, 81
SyQuest cartridges, 108–110
systems, calibrating. *See* calibration

T

tagged image file format (TIFF), 92–93, 95. *See*
    also file formats
tape, storing images on, 111–112

testing scanners, 34

thermal wax printers, 11

third-party software, 38–39

threshold, 171–172

TIFF file format, 92–93, 95. *See also* file formats

time considerations, 4, 34–36

transmissiveness, 29, 124

transparencies, tone range and, 32

transparency option scanners, 24–26

TWAIN, 38

U

under color additions (UCA), 86

under color removal (UCR), 84–85

unsharp masking, 69, 71

V

video, 11, 122

viewing booths, 57

virtual memory, storing images in, 108

W

web presses, 10

Z

zooming, 65

Ziff-Davis Press Survey of Readers

Please help us in our effort to produce the best books on personal computing.
For your assistance, we would be pleased to send you a FREE catalog
featuring the complete line of Ziff-Davis Press books.

1. How did you first learn about this book?

Recommended by a friend . . . . ☐ -1 (5)
Recommended by store personnel . . . . ☐ -2
Saw in Ziff-Davis Press catalog . . . . ☐ -3
Received advertisement in the mail . . . . ☐ -4
Saw the book on bookshelf at store . . . . ☐ -5
Read book review in: _____ ☐ -6
Saw an advertisement in: _____ ☐ -7
Other (Please specify): _____ ☐ -8

2. Which THREE of the following factors most influenced your decision to purchase this book? (Please check up to THREE.)

Front or back cover information on book . . . ☐ -1 (6)
Logo of magazine affiliated with book . . . . . . ☐ -2
Special approach to the content . . . . ☐ -3
Completeness of content . . . . ☐ -4
Author's reputation. ☐ -5
Publisher's reputation . . . . ☐ -6
Book cover design or layout . . . . ☐ -7
Index or table of contents of book . . . . ☐ -8
Price of book . . . . ☐ -9
Special effects, graphics, illustrations . . . . . . ☐ -0
Other (Please specify): _____ ☐ -x

3. How many computer books have you purchased in the last six months? _____ (7-10)

4. On a scale of 1 to 5, where 5 is excellent, 4 is above average, 3 is average, 2 is below average, and 1 is poor, please rate each of the following aspects of this book below. (Please circle your answer.)

Depth/completeness of coverage   5 4 3 2 1 (11)
Organization of material   5 4 3 2 1 (12)
Ease of finding topic   5 4 3 2 1 (13)
Special features/time saving tips   5 4 3 2 1 (14)
Appropriate level of writing   5 4 3 2 1 (15)
Usefulness of table of contents   5 4 3 2 1 (16)
Usefulness of index   5 4 3 2 1 (17)
Usefulness of accompanying disk   5 4 3 2 1 (18)
Usefulness of illustrations/graphics 5 4 3 2 1 (19)
Cover design and attractiveness   5 4 3 2 1 (20)
Overall design and layout of book   5 4 3 2 1 (21)
Overall satisfaction with book   5 4 3 2 1 (22)

5. Which of the following computer publications do you read regularly; that is, 3 out of 4 issues?

Byte. . . . . ☐ -1 (23)
Computer Shopper . . . . ☐ -2
Home Office Computing . . . . ☐ -3
Dr. Dobb's Journal . . . . ☐ -4
LAN Magazine . . . . ☐ -5
MacWEEK . . . . ☐ -6
MacUser . . . . ☐ -7
PC Computing . . . . ☐ -8
PC Magazine . . . . ☐ -9
PC WEEK . . . . ☐ -0
Windows Sources . . . . ☐ -x
Other (Please specify): _____ ☐ -y

Please turn page.

6. What is your level of experience with personal computers? With the subject of this book?

	With PCs	With subject of book
Beginner	☐ -1 (24)	☐ -1 (25)
Intermediate	☐ -2	☐ -2
Advanced	☐ -3	☐ -3

7. Which of the following best describes your job title?

Officer (CEO/President/VP/owner) ☐ -1 (26)
Director/head ☐ -2
Manager/supervisor ☐ -3
Administration/staff ☐ -4
Teacher/educator/trainer ☐ -5
Lawyer/doctor/medical professional ☐ -6
Engineer/technician ☐ -7
Consultant ☐ -8
Not employed/student/retired ☐ -9
Other (Please specify): _____ ☐ -0

8. What is your age?

Under 20 ☐ -1 (27)
21-29 ☐ -2
30-39 ☐ -3
40-49 ☐ -4
50-59 ☐ -5
60 or over ☐ -6

9. Are you:

Male ☐ -1 (28)
Female ☐ -2

Thank you for your assistance with this important information! Please write your address below to receive our free catalog.

Name: _____

Address: _____

City/State/Zip: _____

Fold here to mail.

2974-17-04

_____
_____
_____

BUSINESS REPLY MAIL

FIRST CLASS MAIL   PERMIT NO. 1612   OAKLAND, CA

POSTAGE WILL BE PAID BY ADDRESSEE

Ziff-Davis Press
5903 Christie Avenue
Emeryville, CA 94608-1925
Attn: Marketing

ZIFF-DAVIS
ZD
PRESS

NO POSTAGE
NECESSARY
IF MAILED IN
THE UNITED
STATES